Christoph von Mettenheim

EINSTEIN, POPPER AND THE CRISIS
OF THEORETICAL PHYSICS

Christoph von Mettenheim

EINSTEIN, POPPER AND THE CRISIS

OF THEORETICAL PHYSICS

A NEW APPROACH TO AN ANCIENT PROBLEM

To the Memory of
Sir Karl Raimund Popper (1902–1994)

TABLE OF CONTENTS

INTRODUCTION:
THE ISSUE AT STAKE

'It is not the barrier, which thinking sets us that we want to throw down, but the barrier, which our senses set us.'

HEINRICH HERTZ

The situation of theoretical physics in the 20th Century; great theories developed at the beginning; great problems remaining open at the turn. I. Gravity unexplained; approaches to the problem. II. Background of ideas needed for satisfying both physicists and non-physicists; Einstein's mathematical self-contradiction. III. Fairness to Einstein, gratitude to Popper.

The development of theoretical physics in the first decades of the 20th Century was spectacular. In 1900 Max Planck presented his quantum theory which was to dominate the century. Five years later, in his so-called *annus mirabilis,* Albert Einstein published his light quantum hypothesis, his special theory of relativity and several other papers, among them that on the equivalence of matter and energy, containing (in a different notation) his famous formula $e = mc^2$. In 1911 Ernest Rutherford introduced the theory of the planetary model of the atom and Charles Wilson made traces of subatomic particles visible in cloud chambers, providing atomic theory thus with an experimental outlook. In 1913 Niels Bohr integrated the planetary model into quantum theory and thereby initiated quantum mechanics. And in 1916 Einstein published the foundation of his general theory of relativity and soon began to work on an even more general theory that was to unite the formulae of the electromagnetic field with those of the gravitational field. All this happened in less than twenty years.

Based on those developments other theoretical physicists described more and more particulars of the structure of the atom and its nucleus in the following decades. The planetary model of the atom gradually developed into the *Standard Model of Particle Physics*, a model-theory assuming atoms to consist of various families of particles with different properties, some of them within the nucleus, others orbiting it in the manner of planets orbiting the sun.

In spite of those developments the end of the century left theoretical physics in a difficult situation because some of its greatest problems had remained unresolved. The missing Higgs particle raised one of them. In 1964 the British physicist Peter Higgs and the Belgian physicist François Englert had shown that finalizing the standard model of particle physics required the existence of one more particle[1]. They had even specified the physical properties it would have to own, but by the end of the century no physicist had been able to find it.

Another problem left open was gravity. Popular view assumes it to be a force of attraction emanating from matter itself. Most theoretical physicists however will agree that this merely obscures the issue. As long as we cannot identify the agent by which the supposed attracting force of matter is acting on other material bodies this 'explanation' would be tantamount to admitting in physics miracles in the guise of action at a distance. Physical bodies would be influencing one another without a physical connection between them. Newton had already seen that but left the problem unresolved. Einstein had tried to solve it by introducing the concept of 'curved space' in his General Theory of Relativity. But although most theoretical physicists accepted his theory in principle they still considered the cause of gravity unexplained.

---

1 François Englert/Robert Brout, *Broken Symmetry and the Mass of Gauge Vector Mesons*, Phys. Rev. Lett. Vol. 13 (1964), p. 321–323; Peter Higgs, *Broken symmetries, massless particles and gauge fields*, Phys. Rev. Lett. Vol. 12 (1964), p. 132; *Broken symmetries and the masses of gauge bosons*, Phys. Rev. Lett. Vol. 13 (1964), p. 508.

The new millennium then brought new developments. In 2011 scientists of the CERN at Geneva reported of neutrino experiments in which *superluminal* velocities had been observed for the first time in history. And in 2012 they identified a particle meeting all the specifications postulated by Englert and Higgs and named it 'Higgs Boson'. The standard model of particle physics was complete and in 2013 Englert and Higgs received the Nobel Prize in physics for their theory.

Anyone expecting those developments to open approaches also to other problems or even to new discoveries was to be disappointed however. The neutrino experiments at the CERN continued until they no longer showed superluminal velocities, then ended abruptly. And the discovery of the Higgs boson did not contribute to explaining anything, not even gravity.

That, in briefest outline, is the problem situation with which this essay will deal. The following remarks are to give readers a first idea of the approach it takes.

I

Having model-theories in science is not self-evident because they can provide explanations only within the scope of their model whereas anything going beyond that would mean that the model itself could not be a true description of reality. That is why model-theories tend to hamper new ideas. This psychological mechanism became one of the main hindrances to the progress of theoretical physics in the 20th Century.

Theoretical physicists usually admit that the standard model of particle physics leaves gravity unexplained. Yet although they can hardly believe to know much about nature as long as they do not even understand the most obvious of its forces, they cling unswervingly to that model. Some argue that gravity is but a weak force compared to those working *within* the atom. They insist on considering the standard model a true description of reality although explaining gravity might require giving up just that description. That will be one of the issues of the essay. If no one has found a plausible explanation of gravity in the more than three centuries since Newton's *Philosophiae Naturalis Principia Mathematica* (1687), then we may need to find a new approach to the problem.

The normal way of getting out of a tangle of that sort would be to start from the problem, stating it as clearly possible and framing the questions fundamental for solving it. In any academic discipline a clear statement of a problem will mostly turn out to be half of the way towards its solution.

The present situation in theoretical physics is more intricate because the background of ideas against which physicists see its problems changed significantly during the past century. At the turn of the 19th Century almost every physicist still believed the whole universe to be filled with an invisible substance called 'ether'. The *ether hypothesis* served as an explanation for various physical phenomena and most of the physical theories developed in the 19th Century relied on it. In the early decades of the 20th Century other theories, such as quantum theory, the theory of relativity and the theory of the planetary model of the atom, superseded it. In our days few seem to remember that ether theory ever existed. For understanding the present situation we must therefore consider also the reasons of that development and their bearing on the problems remaining unresolved.

Investigating the history of a problem before coming to the problem itself will often be necessary in this essay. Quantum theory and the theory of relativity have dominated theoretical physics for so long that every theoretical physicist now living must have been educated in their tradition. Many will find it difficult to look at problems of science from a different angle. Non-physicists may not be influenced by that tradition so strongly but they will often lack the background knowledge needed for discussing problems of theoretical physics.

The long standing of that tradition makes it doubtful whether theoretical physics would ever be able to get out of the present deadlock without help from outside. It needs someone taking an independent view. Providing that independent view is my aim in this essay and it also is why I must try to give both sides their due. Physicists must get a sound explanation of the reasons for looking at problems of theoretical physics from a different angle than their tradition taught them, and non-physicists or students of theoretical physics in early stages of their academic education must also get the background information needed for understanding the present situation. Though not a physicist myself, I am trying hard to satisfy the needs of both camps.

II

I can see no better way of dealing with the situation of theoretical physics in the 20th Century than by showing the contrast between the worlds of thought of two of its most famous thinkers, Albert Einstein and Sir Karl Popper, both of them deeply concerned about the problems of theoretical physics. To give readers a first idea of that contrast I begin by demonstrating two of its features which will be important in the essay.

(*1*) An important feature of Popper's world of thought was in the principle of *methodological nominalism* which he he first explained in his book *The Open Society and its Enemies* (1945)[2]. The following quotation from that book shows it. He wrote there:

> 'While we may say that the essentialist interpretation reads a definition 'normally' from the left to the right, we can say that *a definition, as it is normally used in modern science, must be read back to front, or from the right to the left;* for it starts from the defining formula, and asks for a short label to it. Thus, the scientific view of the definition "A puppy is a young dog" would be that it is an answer to the question. "*What shall we call* a young dog?" rather than an answer to the question "What is a puppy?". (Questions

---

2 Karl Popper, *The Open Society and its Enemies*, vol. 2, Chapter 11, section II (pp. 9–21). For a better understanding of my book I strongly recommend reading at least those twelve pages of Popper's book. The quotation following in my text is from p. 14.

like "*What* is life?" or "*What* is gravity?" do not play any rôle in science.) The scientific use of definitions, characterized by the approach "from the right to the left ", may be called its *nominalist* interpretation, as opposed to its Aristotelian or *essentialist* interpretation.' (Popper's italics).

At this point the words 'description' and 'explanation' may serve as examples for demonstrating the difference between those approaches because distinguishing their meaning will be important in the further course of the essay. In common language those words have different meanings. Einstein and other theoretical physicists did not always distinguish them clearly however. Clarifying the meaning which I will connect with those words in the essay may therefore help to prevent later misunderstandings.

When I use the word 'description', I intend it to stand for repetitions or generalizations of information already known. That agrees with its use in common language. We use descriptions for passing on to others the information we have about some object, or for enabling them to identify that object. But we cannot describe something we do not know. That is why we do not expect from a description any information going beyond the object itself. Even if we generalize the information we are passing on, for instance by saying '*all* swans are white', that would still not be considered an explanation of their colour but only as a (wrongly) generalized description of swans.

The word 'explanation', on the other hand, I will use for situations in which we are being told something new, perhaps even interesting. Putting it more precisely, it will stand for sentences or sets of sentences in which the information contained in the *explicans* (the explaining words or sentences) goes beyond a mere generalization of the information contained in the *explicandum* (the words or sentences describing the state of affairs to be explained). That agrees also with Popper's use of the term 'discovery' which he explained like this:

'We may take it, as a rule, that the *explicandum* is more or less well known to be true, or assumed to be so known. For there is little point in asking for an explanation for a state of affairs which may turn out to be entirely imaginary. (...) The *explicans,* on the other hand, which is the object of our search, will as a rule not be known: it will have to be discovered. Thus, scientific explanation, whenever it is a discovery, will be *the explanation of the known by the unknown.*'[3] (Popper's italics).

Going by that understanding, model theories in science could hardly lead to new discoveries because instead of explaining the known by the unknown, they would have to explain the unknown by the known.

(2) An important feature of Einstein's world of thought was directly opposed to the principle of methodological nominalism. He believed that enquiring into the *meaning,* or the *essence,* of concepts might yield interesting results. A passage from his book *Relativity, the Special and the General Theory* is symptomatic. He wrote there:

'At this juncture the theory of relativity entered the arena. As a result of *an analysis of the physical conceptions of time and space,* it became evident that in reality there is not the least incompatibility between the principle of relativity and the law of propagation of light, and that by systematically holding fast to both these laws a logically rigid theory could be arrived at.'[4] (My italics).

This fragment shows that Einstein considerd the 'analysis of ... physical conceptions' an important topic. In Popper's categories quoted above, Einstein thought that a definition must be read 'from left to right'. And he therefore believed that for understanding time he must answer the question 'what is time?'

---

3 Karl Popper, *Realism and the Aim of Science* (1983), p. 132.

4 Einstein, *Relativity, the Special and the General Theory,* translated by Robert W. Lawson (end of section 7).

The danger of that approach lies in the fact that opinions on the meaning of a word can vary even if the word remains the same. The approach thus opens the possibility of ostensively solving a problem by using some concept in one meaning at the beginning of an investigation and using it in a different meaning at its end without changing the concept itself. We will see that Einstein repeatedly reached his results by that method which I will discuss in detail in *Chapter 9*. At this point an example from his mathematics may serve to show its consequences. Getting the full meaning of his formulae is not important yet but I will explain in footnotes the symbols he used. The only mathematical knowledge needed for understanding the following is that if we do something on one side of an equation, then we must do the same also on the other side.

In his famous paper *On the Electrodynamics of Moving Bodies*[5], containig the first presentation of his Special Theory of Relativity, Einstein violated that principle. After introducing the premise that the speed of light is always constant, independent of the motion of its source, he considered a system with the ends A and B and defined the synchronism of two 'clocks' located at those points by equation

$$t_B - t_A = t'_A - t_B .$$

(1)[6]

The purport of equation (1) is that assuming the speed of light to be constant, a light signal travelling in a system *at rest* from A to B and back to A will take equal time on both ways. Einstein then continued by defining in equation

$$\frac{2\,\overline{AB}}{t'_A - t_B} = V$$

(2)

---

5 The English translation I use is from *The Principle of Relativity*, published by Methuen & Co. (1923).

6 In equation (1) t_A stands for 'time at A', t_B for 'time at B', and t'_A for 'time at A after reflection of light in B'.

the speed of light (V) by the time taken by a light signal on its ways from *A* to *B and back to A.*

Two pages further down, in § 2 of his paper, he introduced the following equations for defining the time taken by light in a *moving* system on its single ways from *A* to *B* and back from *B* to *A* after reflection in *B*[7]:

$$t_B - t_A = \frac{r_{AB}}{V - v} \tag{3}$$

$$t'_A - t_B = \frac{r_{AB}}{V + v} \tag{4}$$

The expressions on the left sides of equations (3) and (4) are the same as those on either sides of (1). Einstein did not introduce new definitions of his symbols. Equation (1) therefore implies that the left sides of (3) and (4) are equal. Their right sides must then also be equal which gives us

$$\frac{r_{AB}}{V - v} = \frac{r_{AB}}{V + v} \tag{5}$$

One glance will show that equation (5) cannot possibly be correct. The numerators on both sides are identical whereas the denominators differ *only* in the symbols '+' and '-'. We can prove this incorrectness mathematically by multiplying each of the numerators with both denominators and then eliminating r_{AB} on both sides, which reduces (5) to

$$V + v = V - v \quad \text{or} \quad + v = - v \quad \text{or} \quad + = - \tag{5a,b,c}$$
(plus equals minus).

---

7 In equations (2) and (3) r_{AB} stands for 'length of the moving system', V for 'velocity of light' (which Einstein assumed to be constant, independent of the motion of its source), and v for 'speed of the moving system'. In a footnote on p. 896 Einstein stated explicitly that the meaning of t in (2) and (3) *was to be the same as in* (1). In English translation the wording of the footnote is '*Time here* stands for *time of the system at rest* and also for the *position of the hands of the moving clock which is at the place under discussion*'. I will discuss it in *Chapter 9, IV.*

This shows that Einstein's equations either violate the definition of '=' or, alternatively, one of those of '+' or '-'. The origin of that self-contradiction lay in equation (2), where he shifted the meaning of one of the symbols he used. We will see in *Chapter 9, IV* that it was caused by transferring to the *moving* system the definition of *V* which he had introduced in equation (2) only for a system *at rest*.

(3) The fact that a mistake of that importance could have remained un-discovered so long is more interesting than the mistake itself. It casts a strange light on the state of theoretical physics in the past century. Even if Einstein's self-contradiction were explicable somehow, one still would expect some other physicist to have discussed or even published that explanation somewhere. But that never happened. It was one of my reasons for writing *Albert Einstein oder Der Irrtum eines Jahrhunderts* (2009) of which this essay partly is a translation.

My main reason was different however. Some years earlier I had found what I believe to be a plausible explanation of gravity and had published it in my book *Popper versus Einstein* (1998). It means returning to ether theory and it is incompatible with Einstein's theory of relativity. I could not avoid criticizing that theory therefore, and did so with arguments not based on mathematics but on logic alone. Although I had written in English that book no theoretical physicist ever took notice of my criticism, at least not to my knowledge. And none considered the explanation of gravity which I proposed there to deserve discussion. That seemed to indicate that I had overrated the power of logic in science and had underrated the power of inertia created by a long tradition.

For overcoming that problem I had to find arguments not only at the level of logic but also at that of mathematics. The fact that Einstein had based his special theory of relativity on the mathematical self-contradiction shown above gave me a new argument supporting my own explanation of gravity. And the fact that his mistake had remained undiscovered in a whole century gave me a new argument for demonstrating the weakness of the tradition of theoretical physics in the 20th Century. If a mathematical mistake as obvious as the one just shown could remain undiscovered for a whole century, then what are we to think of the far more complicated calculations of quantum theory or of the general theory of relativity?

My new book *Albert Einstein oder: Der Irrtum eines Jahrhunderts* did not improve the situation however, possibly because one of my books now was in English and the other in German. I published a short English paper discussing only Einstein's mathematical mistake but never heard of any reactions to it[8]. I can only put that down to a feeling of embarrassment. If anybody could meet my mathematical argument, then surely some theoretical physicist would have done so by now. It thus became necessary to provide an English translation of the German book and to integrate in it my explanation of gravity which I will state in *Chapters 11* and *12*. That is the purpose of this essay. For achieving that aim, I had to change the order of discussion in some parts of this book. I also tried to improve some of the arguments stated in the German version. In the end this book had to be given a new title in order to show that it is more than a mere translation.

Earlier versions of this essay were offered to all the important publishers of theoretical physics or of philosophy that I could make out. All of them refused it. Self-publishing was the only way left open to me for trying to help theoretical physics in getting out of its present deadlock. I can only hope that the book will find readers willing to persevere to its end, and sufficiently open-minded for figuring out themselves the reasons for the attitude of those publishers.

III

Though criticizing Einstein severely in this essay, I am doing my utmost to be fair to his memory. By neither styling him a genius nor accusing

---

8 C.v. Mettenheim, *The Oscillation Project with Emulsion-Tracking Apparatus (OPERA) experiment: An argument for Superluminal Velocities?*, Physics Essays, vol. 25 (2012), p. 397–403.

him of dishonesty I think I am not only doing him more justice but also being fairer to him than his uncritical admirers are. All those following him blindly put on him alone the responsibility for leading theoretical physics astray in more than a century. They fail to see that the far more important reason for that crisis is not in Einstein but in his followers. It is in their lack of critical faculties and of independency of thinking, in their exaggerated desire for geniality and in the barren intellectual soil which that attitude left over for permitting creativity to survive also in the field of science.

Einstein's real achievement remained almost unnoticed over all that adulation. His contribution to the theory of scientific discovery, which I will discuss in *Chapters 4* and *5*, was truly revolutionary but it was so original that few theoretical physicists seem to understand it to this day. That was his personal tragedy.

Karl Popper was the one to see that most clearly. My admiration and my personal debt of gratitude belong to his memory and friendship first. By not stopping at criticizing even him, not for his theory of scientific discovery but sometimes for his way of applying it to theoretical physics, I am trying to pay back that debt in the way he would have wanted.

ON EINSTEIN'S THEORY OF KNOWLEDGE

> 'The difference between Kant and Einstein is not in the fact that one of them assumed Euclidean space while the other assumed non-Euclidean space but most of all in the relation, which they established between mathematics and reality.'
>
> FRIEDRICH DÜRRENMATT

Experimental physics came first; theoretical physics initially was only stopgap; beginning of secession; stagnation of theoretical physics in 20ᵗʰ Century; problem of gravity unresolved. I. The spirit of Enlightenment encouraged speculative science. II. Contrast with timid undercurrent pining for certainty. III. Einstein's ambivalent attitude; overrating mathematics symptomatic of the period; slow liberation from Aristotelianism; conditions of Einstein's success; emergence of 'the Schism in Physics'.

Experimental physics is as old as physics itself. Anyone throwing a stone at a target is performing a physical experiment. We can therefore be certain that physical experiments of the most simple kind were carried out since man began to have thoughts about nature. And ever since he handled fire and water he must have consciously targeted such experiments.

Theoretical physics, by contrast, seems to have emerged only in the 19th Century, and developed only gradually even then. At present its representatives consider it a largely autonomous branch of science, and that development is interesting not only for the history of science but also linked with the beginning of industrialization and hence with European economic history.

Technical inventions often foster the progress of humanity, and inventors were appreciated at all times. Since the end of the Middle Ages many countries specially protected them. The Venetian Republic made first attempts at creating a kind of patent law in the 15th Century already. Industrial development and the French Revolution then encouraged the general recognition of intellectual property eventually to be recognized worldwide. Inventions increasingly often became sources of personal wealth, waking creative powers thereby and unleashing a kind of intellectual gold rush in the 19th Century. The focus of economic life, formerly determined mainly by agriculture and tade, shifted towards commercial production which was beginning to to develop into industrial production. Internationally, the manifold competition of individuals resulted in bitter economic strife of whole nations. On the European continent the foremost aim was catching up with the technical lead which British industry had gained by James Watt's invention of the steam engine (1765) and Edmund Cartwright's of the power loom (1785), and which Great Britain jealously defended.

That economic development generated an unprecedented demand for physicists. Many felt the future of mankind to be lying in technology and engineering, and entrusted to them their own future. The need for physicists entailed that for physical education which also took a steep rise in the 19th century. Chairs for teaching physics only in theory were mere stopgaps at first, accepted of necessity because colleges and universities were unable to meet demand otherwise. Teachers were available, even interested and talented teachers endowed with knowledge, enthusiasm and deep understanding of physical phenomena. Nor did classrooms cause any problems. Only laboratories for performing experiments were scarce because they required financial backing. Due to that dilemma and in order to let young physicists at least have theoretical training, professors were appointed even where laboratories were not available. Some teachers of physics had to wait for years until at last they could carry out their own experiments under tolerable circumstances. Heinrich Hertz, the discoverer of electromagnetic waves, was a famous example[9].

---

9   Albrecht Fölsing, *Heinrich Hertz - Eine Biographie*, Parts I–III.

Markets follow their own laws however, and the market for physical knowledge soon began to generate its own momentum. Teachers of physics without a laboratory of their own would try to remedy that unsatisfactory state of affairs by attracting the more attention to their theoretical teachings and publications, hoping these might procure them a better-equipped chair elsewhere. Initially less-favoured teachers thus were encouraged to excel in the field of theory, and their achievements would attract students bent on theory rather than on practice, gifted rhetorically but preferring to leave experimenting to others. The most capable of them would often become academic teachers themselves later. The officials appointing them would usually have but limited knowledge of physics, and would depend for their selection on publications they understood even less, or on experts who had to be physicists if they were to be qualified. Thus theoretical physics gradually came to be generating itself. In the second half of the 19th Century first chairs were instituted for teaching physics in theory only. Hendrik Antoon Lorentz (1853–1928) and Max Planck (1858–1947) were among the first physicists to be pure theorists from the outset[10].

By this process unfolding in the 19th Century, theoretical physics gradually established itself as a separate branch of science, largely independent of experimental physics. Since then its home has come to be somewhere in no-man's-land between physics, mathematics and philosophy. Its relations to those disciplines remind of those of the newborn baby to the fairies in the tale. Physics endows it with the subjective certitude of empirical knowledge. Mathematics presents it with the objective certainty of logical reasoning. And philosophy is the evil fairy coming uninvited. As her christening present she puts in the cradle of the child the question of how the subjective certitude of empirical knowledge based on experience and the objective certainty gained by the rules of logic and mathematics together can yield knowledge that is certain yet extends beyond experience. And along with that she presents it with the eternal riddle of how human knowledge is possible at all.

---

10 Karl v.Meÿenn in *Die Großen Physiker*, vol. II p. 87, 89; Fölsing, *Heinrich Hertz – Eine Biographie*, p. 195.

Physical science shifted its focus considerably in the course of that development. In the 19th Century experimental findings, such as Faraday's discovery of electrical induction, Young's of the polarization of light, Roentgen's of x-rays, Becquerel's of radioactivity, or Hertz' of electromagnetic waves, stood in the foreground. Since the turn of the century theoretical physics began to attract more attention. The development of quantum theory, the theory of relativity and the Rutherford-Bohr planetary model of the atom all came in barely fifteen years. They did not affect daily life in any way, but they changed the worldview of modern physics more than any other discovery had changed it since the days of Copernicus. The quest for theoretical unification of physics began to rouse more interest than discoveries of physical effects. Parallel to that, theoretical physics increasingly withdrew from physical practice and from the understanding of common people.

At present however, theoretical physics has been stuck in a crisis for decades. No fundamental advance in theoretical physical knowledge has been on record for a long time. The impulses to technological progress come from other disciplines. Yet business ostensibly goes on as usual. Black holes, or the discovery of new or still smaller particles, or some other news of that kind will occasionally be reported. Quantum computers have been predicted for decades but never been put in practice. The effect of such reports coming up with almost somniferous regularity has long been that the public will hardly take notice of them anymore because common understanding of theoretical physics has suffered too long.

In spite of that however, credulous politicians continue investing in theoretical physics public funds of dimensions hardly conceivable because they, too, are unable to follow its ways[11]. They depend for their decisions on expert opinions from theoretical physicists or on publications in journals of science staffed also with theoretical physicists. At present theoretical physics not only generates itself. It also controls itself.

---

11 The text refers to the development of large particle colliders. At the outset of those experiments, many hoped they might lead to making nuclear fusion useful for civil purposes. At present there seems to be no question of that anymore.

Results are nowhere in sight however. The most fundamental questions of physics remain unanswered. That applies not only to controlled thermonuclear fusion which seemed to be almost within reach so long but now is removed again to a distant future after having devoured billions of money in any currency[12].

It also applies to the relationship of light and matter and to the explanation of gravity. In fact, it applies to all the truly big problems of theoretical physics and astronomy, even to explaining the causes of the earth's rotation or of magnetism. Not even new approaches to their solution were found in so many decades. They appear so hopeless that no one seems willing to take them on anymore. Non-physicists hardly remember that those problems still exist, and even physicists seem sometimes to have forgotten them. Regarding the complexity of issues left open, the current state of theoretical physics is more hopeless than that of the ancient Ptolemaic theory had been in the times of Copernicus.

I

There is a link between the development just described and the history of thought in Europe. The stunning progress of experimental science in the 19th Century had been a belated effect of the independent and critical spirit of the Age of Enlightenment. Great developmens at the close of the Middle Ages had wakened that spirit, foremost the discovery of America in 1492 and the heliocentric theory put forward by Nicolaus Copernicus (1473–1543) about in 1514. They did not immediately influence the worldview of individuals but in the further course of history they changed the worldview of humanity radically and forever. They opened to human

---

12 Constructing the fusion reactor in southern France began only after I had published *Albert Einstein oder Der Irrtum eines Jahrhunderts* on the internet in 2005. Details of the experiment are on the ITER website. At the time of writing this book the estimate of total costs is about 13 billion Euros but its outcome still is open. I do not expect it to succeed, and will explain my reasons for that in *Chapter 11, VI*.

imagination ranges of dimensions hitherto quite unknown, here on earth as well as in the universe. And the art of printing developed by Johannes Gutenberg (1395–1468) around 1455 lent to that imagination the wings on which the minds of countless individuals could freely tour those vast spaces.

That fascinating enlargement of perspectives began to take effect in the Age of Enlightenment. In science Galileo Galilei (1564–1642), René Descartes (1596–1650), Christiaan Huygens (1629–1695) and Sir Isaac Newton (1643–1727) were among the first to succeed in throwing off the mental shackles of the past. By the end of the 18th Century their thoughts had circulated widely and had generated a spirit of optimism carrying many with it. No method was prescribed or prohibited. One pondered over the mysteries of nature and speculated and experimented at heart's content.

Success was all that counted, and it strongly took sides for that free, undogmatic and open science, particularly in physics. All the great discoveries about electricity, from Galvani's first studies on animal electricity (1789) to Faraday's discovery of electrical induction (1831), came in barely more than forty years. The discoveries of light interference and of the polarization of light also fell within that period. And events almost toppled over themselves once scientists had realised that behind the world of visible phenomena there was another world of invisible phenomena still awaiting discovery.

Cathode rays, X-rays, radioactivity and electromagnetic waves, the photoelectric effect and many more physical phenomena normally hidden by nature to human perception were discovered within a few decades. At the turn of the 19th Century the world of experimental physics was upside down, and physical science had to begin almost from scratch in some fields.

Practical application did not tarry either. Great explorers like Galvani, Young, Faraday, Ampère, Hertz and Roentgen had shown the way, and great technicians such as Watt, Cartwright, Stephenson, Morse, Siemens, Edison and Otto followed them hotfoot. Their epoch-making inventions of the steam engine, the power loom, the locomotive, the telegraph, the dynamo, the light bulb or the combustion engine, to mention only some of the wonderful achievements of that period, changed the world in few decades to an extent that previously found no comparison in millennia.

II

Parallel to that development, however, there lived in the world of science also a more timid undercurrent of thought, finding it more difficult to own up to new discoveries. Its exponents could not quite muster the courage needed for making experiments, but rather would squint at the opinions of others. Satisfying curiosity by experimenting is a risky thing after all, especially for scientists pining for public attention.

It takes readiness to commit mistakes, to confess one's own ignorance when they become visible, and to admit to oneself and others that one had been wrong. Only strong natures could face that; others feared loss of reputation. They tried to avoid the risk of error, strove for infallibility, and clung to the certainty of mathematics. Like the amanuensis Wagner in Goethe's *Faust* they believed they knew much, and hoped to know everything one day.

Those following such lines of thinking would be inclined to place knowledge above curiosity. They would be less interested in making new discoveries, and averse to shaping opinions of their own. They would also tend to be impressed by rhetoric more than by creativity, even if paired with scientific discipline. Theorists were exposed to those dangers more than experimentalists were, whom nature itself would permanently demonstrate in their experiments the limits of their knowledge.

On the European continent the admired Anglo-Saxon example did the rest. In late 19th Century the mathematical reflections of the Scottish physicist James Clerk Maxwell (1831–1879) probably were the most widely accepted doctrine of physics[13]. At the end of the century that school

---

13 Maxwell's *Treatise on Electricity and Magnetism* (1873) and his *Matter and Motion*

prevailed in theoretical physics, and reinforced the above-mentioned development favouring too long and too strongly physicists interested in theory, gifted mathematically and rhetorically, but turned away, rather, from experimental practice.

The *First Part* of this book will mainly be about the clash between those two major intellectual streams which I have tried to outline. It probably is as old as science itself, being the expression of a fundamental conflict that becomes visible whenever people consciously turn their minds to the world in which we live, and try to understand it. The contrast between Faust and his amanuensis Wagner is everlasting. In some respects it also describes that between Socrates and Plato, or between Huygens and Newton, or between Faraday and Maxwell [14]. No science is spared that conflict between critical and dogmatic thinking because every scientist must make it out with himself, in each generation, and not only once in his life but throughout his scientific work. Some go boldly ahead, devoting their lives to their passion for truth and taking on greatest risks for its sake. If they fail they will soon be forgotten. If they succeed however, and if their success becomes visible, then others will emulate them and sometimes even betray them for their own advantage.

(1877) probably were the most comprehensive statements of physical mathematics of that time. Einstein took Maxwell's theory as a starting point in his papers *Zur Elektrodynamik bewegter Körper* (1905, On the Electrodynamics of Moving Bodies) and *Ist die Trägheit eines Körpers von seinem Energieinhalt abhängig?* (1905, Does the Inertia of a Body Depend on its Energy Content?).

14 For the difference between the attitudes of Socrates and Plato, see Karl Popper's *Open Society* vol. I, *The Spell of Plato, Chapter 10*, sections *V – VII*. The differences between those of Huygens and Newton or of Faraday and Maxwell will become apparent in the further course of this essay.

III

The contrast between critical and dogmatic thinking is alive also in theoretical physics, and it shows in Einstein's oeuvre particularly. Einstein not only embodied it in his own person like no other physicist; he even realised it in his life and in his work. Two souls were living in his chest, and trying to separate ever and again. Light and shadow were so close together in his thinking that the borderline between them sometimes hardly is traceable. That is why Einstein's world of thought is the framework of this essay whereas Popper's mehodology is its guideline. Einstein's physical theories and his reflections on the theory of science were a mirror of his time. Their influence on contemporary theoretical physics can hardly be overrated.

We will see that despite the exceptional status usually accorded him, Einstein was not an exceptional case. In fact, rather on the contrary, he was very normal, almost too normal. Being interested not only in his proper field of physics but beyond that also in philosophy and in the theory of science, he was in some respects the model scientist of his time. That was why his contemporaries could so easily believe him to be a genius. Many shared his notion of science, and his greatest problem was that this notion itself aimed at a conflict which he first made visible without being able to resolve it.

Einstein wanted to be an empirical scientist as we will see. But he also wanted to have the certainty of mathematics on his side. That was why he was always in line with the most progressive scientists of his time where he was right. And it also explains why he was mostly following a line of thought far older than the conditions from which he started where he was wrong. His errors were almost invariably expressions of a tradition of many centuries, influencing not only theoretical physics. In fact, Einstein's unfortunate pioneering success consisted in importing into physical theory the intellectual approach of essentialism mentioned in section *II* of the *Introduction*, which in philosophy goes back as far as to Plato and Aristotle.

In that, too, Einstein was in harmony with his time. The approach he followed dominated the late 19th Century and infected many of the theories brought forward by theoretical physics in the past 20th Century. Max Planck also stood under its influence. One of its symptoms was an overestimation of mathematics, caused by disregarding its origin and its limitations. It shows in Planck's original quantum theory as well as in its later developments strongly influenced by Niels Bohr and Werner Heisenberg.

The state of affairs was unsatisfactory but it had also a positive aspect. It attracted the attention of others to the problems of theoretical physics, Poincaré, Tarski, Gödel and Popper among them, and thus eventually resulted in the great developments made in the *theory* of science in the 20th Century – developments which theoretical physicists mainly neglect to this day.

We must begin at the beginning however. For seeing the problem, we must understand the intellectual climate from which Einstein's approach arose. That background is more interesting in some respects than later developments. Logical mistakes happen everywhere, even in science. That is why critical discussion is so important. We will see Max Planck's logical mistakes in *Chapter 8*, but although we can locate them precisely, we will also see that they were by no means obvious. Planck's problem consisted in asking wrong questions rather than giving wrong answers. Einstein's oeuvre raises greater questions. We saw one of his mistakes in the *Introduction* already, and will see more of it in *Chapter 9*. It was so obvious that it seems almost to leap to the eye. Yet, instead of being rejected on the spot, his theories roused astonishment and admiration. The mistakes he made were in principle but harmless blunders that might easily happen to a young scientist. If theoretical physics had been on the alert at the time instead of being mainly on the lookout for geniality, then one of his teachers or some critical reader would have noticed that he was confusing the meaning of his terms and that his deductions were mathematically incorrect. And there the matter would have ended.

The greatest mystery is why that never happened. Why did not only famous contemporaries at home and abroad, Max Planck and Hendrik Antoon Lorentz among them, but almost all physicists worldwide and even the great philosopher Karl Popper accept Einstein's theory of relativity almost unhesitatingly without taking umbrage at its inherent contradictions? We will see in *Chapter 8* that in his theory of the quantum nature of light Einstein shifted the meaning of Max Planck's original quantum theory without even mentioning that. Why did others tolerate that without considering what it entailed? How could it be that ether theory, accepted by most physicists since the times of Huygens and Newton, was given up almost unopposed in favour of a new theory put forward by a hitherto unknown 'expert of third class' of the Swiss Patent Office? How could all this happen although his theories left unexplained so much that ether theory previously had explained? Why were the achievements of generations of physicists set aside so unceremoniously? Is it possible that the relationship to logic of theoretical physicists is generally disturbed, or that they will uncritically adopt opinions of others instead of forming theirs themselves?

Such questions will come to mind. They appear to me more interesting than Einstein's mistakes because they lead to the roots of the problem instead of dealing with its symptoms only. Einstein was a child of his age after all, and his thoughts were with the mainstream of that period. If my interpretation is correct, then the answers to those questions follow mainly from two causes. One of them was in the tradition of epistemology which only gradually liberated itself from the age-old traditions and mental shackles of the Aristotelian school of thought. And the other was in the specific historical situation in which physics stood at the beginning of the 20th Century. We will see several factors interacting then.

How could Einstein's theory of relativity be so overwhelmingly successful? It seems impossible to find an even halfway plausible answer to that question in the originality of the theory or in its power of persuasion. He never was the unrecognized genius pining for being discovered, but was accepted almost from the moment when he first walked onto the stage of science as a young man, and became world famous only a few years later. I think he owed that easy victory not to the strength of his theories but mainly to the historical constellation in which he proposed them. He framed his thoughts at a time when the revolutionary developments of the 19th Century had thrown the self-confidence of many physicists into a deep crisis. His theory of relativity ostensibly showed them a way of overcoming that crisis, and of conceiving physics as an 'exact science' again. He promised to achieve that for which many scientists were yearning. His theories served what they wished for far better than they served the progress of science.

I will explain this view in the following chapters. If my interpretation is correct, then the turn of the 19th Century marks an important parting of the ways in the history of physics. In those decades experimental physics and theoretical physics finally separated. Some scientists continued to search for new physical effects and explanations, inventing original experiments and making wonderul discoveries such as the semiconductor or laser technology which were to shape the technical and economic development of the late 20th century to an almost incredible extent. Others worked on the presentation and perfection of their theoretical *system*. That was the origin of the ,Schism in Physics'[15] described by Karl Popper, which has since been deepening more and more the gulf between theoretical physics and other sciences, and has been leading even within physics itself to problems of understanding hardly to be bridged any more in our time.

The striking stagnation of theoretical physics in the second half of the 20th Century is a direct sequel to that misguided development about a century ago. It originated in a misunderstanding of the principles of science, the symptoms of which show clearly in quantum theory and the theory of relativity, and it still reigns unchallenged in most other areas of theoretical physics. We will see those relationships in detail in the following chapters. For really understanding them however, and for being fair to Einstein's memory, we must begin by taking a closer look at the circumstances of his life.

---

15 The quotation refers to Popper's book *Quantum Theory and the Schism in Physics* (1982).

CHAPTER 1:

EINSTEIN AND THE PROBLEM OF GRAVITATION

Einstein's fame; his influence on the method of theoretical physics. I. Outline of his biography; mysterious decline of creativity; general relativity remaining torso; shipwrecked on general field theory and world formula; gravity left unexplained; attractive force would be action at a distance; Newton's letter to Bentley; gravity gretest challenge to theory of science. II. Causes of Einstein's failure; conditions of discussion disturbed by World War I; Poincaré, Whitehead, Russell, Tarski, Gödel, Popper ignored by theoretical physics.

Albert Einstein was born at Ulm, Germany, on March 14, 1879, and died at Princeton, U.S.A., on April 18, 1955. Many of his contemporaries considered him the greatest physicist of his time. The spell of his personality still lives more than sixty years after his death. His achievements in science, his moral integrity, his kind appearance and his personal modesty probably made him the most admired man of the 20th Century. To this day he remains the epitome of a genius.

Einstein's is fame began with his theory of relativity but it does not stand on that alone. He received the Nobel Prize of 1921 for physics not for that theory but for his theory of the quantum nature of light which he had also published in 1905. His thoughts changed theoretical physics in a way that was downright revolutionary. Even today they still have an influence of which some physicists are barely aware. The Big Bang hypothesis, the theory of 'Black Holes', the laser and the discovery of nuclear fission, all have been linked with his theories[16].

---

16 Except for the theory of Black Holes, that connection does not exist. For the discov-

His influence on the *method* of theoretical physics is less well known than those individual relationships but more important in its general impact. He gave countless lectures, several of them dealing with questions of epistemology, and published them worldwide. With such issues he not only reached a large audience but also influenced the direction of research in theoretical-physics decisively for decades. There is a powerful philosophical dimension to his work.

I

Einstein's philosophical ideas were rooted in his work as a theoretical physicist. Fame came to him so early in life that he could barely escape public interest in his person. His journeys took him all over the globe, and his lectures were widely reported. He carried on correspondences revealing the origin and the development of his thoughts. His every step is traceable as it were, with hardly any exception. Yet in spite of that publicity, understanding his life still troubles his biographers. The curve of creativity in the life of the famous man runs in a strange cadence, reversing almost in order the curve of his fame.

He achieved his most admired results early in life. In 1905, when he published his papers on the light quantum hypothesis, on the Brownian motion and on the special theory of relativity, he was aged only 26 years and, as an 'Expert of Third Class' of the Patent Office of Berne, quite unknown to the world of science[17]. That same year he also developed the

---

ery of nuclear energy I will explain that in detail in *Chapter 5*; for Big Bang-theory see my ‚*Popper* versus *Einstein'* (1998), Chapter 9.

17   Albert Einstein, *Über einen die Erzeugung und Umwandlung des Lichts betreffenden heuristischen Standpunkt* (Concerning an Heuristic Point of View Toward the Emission and Transformation of Light), Annalen der Physik (1905) p. 132 ff.; *Über die von der molekulartheoretischen Theorie geforderte Bewegung von in ruhenden Flüssigkeiten suspendierten Teilchen* (On the Motion of Small Particles Suspended in a Stationary Liquid, as Required by the Molecular Kinetic Theory of Heat*)*, Annalen der Physik (1905) p. 549 ff.; *Zur Elektrodynamik bewegter Körper* (On the Electrodynamics of Moving Bodies), Annalen der Physik (1905) p. 891 ff.

equation $e = mc^2$, today probably his most widely known formula and still considered by many as containing the explanation of nuclear energy which was to be discovered more than thirty years later[18].

Only eleven years after his special theory, which had found widespread recognition by then, he published his paper *On the Foundations of General Relativity*, predicting among other things the curvature of light rays in the gravitational field of the Sun[19]. His world fame began when the British physicist and astronomer Arthur Eddington confirmed that prediction in his reports on the British eclipse expeditions of 1919 and 1922.[20]

Modern theoretical physicists consider Einstein's general theory of relativity as having overcome Newton's theory of gravitation. Some see in it his greatest achievement[21]. It remained a torso however, showing mathematical approaches but permitting neither direct applications nor revealing a general principle for deriving them. In our time theoretical physicists rarely call special relativity into question, but they still treat the general theory controversially, particularly with respect to specific conclusions following from it. Some theorists take an interest in it, but for physical practice it is irrelevant[22]. And that was the last of the great theories Einstein developed. In the field of physics he continued publishing some general and popular texts and some ideas on quantum theory. The next decades of his life, however, he mostly spent in search for a unified field theory that

---

18 Albert Einstein, *Ist die Trägheit eines Körpers von seinem Energieinhalt abhängig?* (Does the Inertia of a Moving Body Depend on its Energy Content), Annalen der Physik (1905) p. 639 ff.; Einstein was using a different notation there.

19 Albert Einstein, *Die Grundlage der Allgemeinen Relativitätstheorie* (The Foundation of the General Theory of Relativity), Annalen der Physik (1916), p. 769 ff.

20 It is open to doubt whether this confirmation was correct. According to Fölsing and Hawking, the results reported by Eddington were not significant but only interpreted by him as confirming the theory of relativity (Albrecht Fölsing, *Albert Einstein - Eine Biographie*, p. 492-500; S. W. Hawking, *A Brief History of Time'* p. 102).

21 See for instance   Max Born, *Die Relativitätstheorie Einsteins* (1920), 7.ed. (2003), p. 271f.; S. W. Hawking, *A Brief History of Time'* p. 33f.

22 Some physicists might disagree with my statement and point to the experiments on time dilation. Even in view of those experiments, I still contend that the general theory of relativity has nothing to do with physical practice, not even with the practice of controlling distant missiles or satellites. I will discuss that in *Chapters 6, IV, 3b* and *9,II*.

was to unite the formulae of the electromagnetic field with those of the gravitational field.

That was where he shipwrecked. Several of his papers were dedicated to the problem. Indeed, sometimes hardly a year passed by in which he did not once again announce, or even publish, its 'final' solution[23]. All those attempts however fell victims either to his own criticism or to that of his contemporaries. The only effect of his numerous failures and premature publications was that his reputation in the professional world took damage, and that scientists no longer heeded his criticism even where it may have been justified[24]. The problem of the unified field theory remained unresolved, and despite serious efforts no other physicist was able to solve it in the many decades since Einstein's death. The impulses to the progress of science, ostensibly originating from the ideas of quantum theory and the theory of relativity in the early years of the 20th Century, gradually and almost unnoticed returned to experimental physics since then. There, in cautiously groping steps, some engineering-based results were reached which sometimes even permitted modest new insights into the mysteries of nature. But no physicist succeeded in finding the 'World Formula' of Einstein's dreams, which would eliminate at a stroke and through a single principle all the inconsistencies of the present tangle. Theoretical physics made no headway in decades.

We might have to put up with that if at least satisfactory solutions had been found to other problems, but there is no question of that. The problem of explaining gravity, which should have been solved if Einstein's General Theory of Relativity had really overcome Newtonian theory, still is unresolved, and that is not even in dispute. Some of the most famous

---

23 Albrecht Fölsing, *Albert Einstein - Eine Biographie,* p. 779 ff; Armin Hermann in v.Meÿenn, *Die Großen Physiker,* vol. II, p. 245f. – Hermann quotes Wolfgang Pauli there as having said that Einstein was bestowing the world with new field theories at an average rate of one per year.

24 The text refers to Einstein's famous criticism of the Copenhagen Interpretation of quantum mechanics (Einstein, Podolsky, Rosen *'Can Quantum-Mechanical Description of Physical Reality Be Considered Complete?,* Physical Review vol. 47 [1935], p. 777–780); see also Karl Popper, *Quantum Theory and the Schism in Physics,* p. 147 ff. For the problem of action at a distance see my *,Popper versus Einstein'* p. 156 ff.

physicists of the past century agreed at least on the fact that the problems of gravitation are still open, and that Einstein's General Theory of Relativity had *not* solved them[25].

That is amazing in itself. More than three centuries have passed by since Newton, in his *'Philosophiae Naturalis Principia Mathematica'* (1687), stated the laws of gravity still recognized for all practical purposes in our days. Yet although his theory proved to be an excellent *description* of gravitation, we still are unable to *explain* the force of gravity[26]. For achieving that, we would need to know what causes attraction and by what mechanism it acts on other material bodies, but not even Newton had been able to discover that. When it comes to explaining gravity we seem to be at our wit's end. We like to think we can know everything, and our mastery of technology even tempts us into believing that we can achieve whatever we try. The barren truth however is that we know not even why we stand on this Earth instead of being catapulted into space at a tangent by the centrifugal force of its rotation according to Newton's first law of motion. Why will Newton's law of inertia, stating that

> 'every body continues in its state of rest, or of uniform motion in a right line unless it is compelled to change that state by forces impressed on it',

be violated by matter itself? What is the mechanism of that mysterious force of gravity, supposedly inherent *in* matter, yet 'impressing' its force *on* matter? Why do its effects still make themselves felt in lightyears of distance? Why can no shielding or filter disrupt them?

---

25 Richard P. Feynman, *QED – The Strange Theory of Light and Matter*, ch. 5; Leon Lederman, *The God Particle*, ch. 3; Stephen Hawking, *A Brief History of Time*, ch. 8. - The Higgs boson recently discovered reputedly is of fundamental importance for the so-called Standard Model of Particle Physics. Even after its discovery that model still cannot explain gravity. I will discuss that in detail in *Chapter 10*.

26 For my distinction between *descriptions* and *explanations* see section *II, 1* of the *Introduction*, above.

Newton himself had seen the problem clearly but left it unresolved. In the *General Scholium* at the end of his *Principia* he wrote:

'Hitherto we have explained he phenomena of the heavens and of our sea by the power of gravity, but have not yet assigned the cause of this power. *This is certain that it must proceed from a cause that penetrates to the very centres of the sun and planets*, without suffering he last diminution of its force; that operates not according to the quantity of the surfaces of the particles upon which it acts (as mechanical forces used to do), but according to the quantity of the solid matter which they contain, and propagates its virtue on all sides to immense distances, decreasing always as the inverse square of the distances. ... But hitherto I have not been able to discover the cause of those properties of gravity from phenomena, and I frame no hypotheses; for whatever is not deduced from the phenomena is to be called an hypothesis; and hypotheses, whether metaphysical or physical, whether of occult qualities or mechanical, have no place in experimental philosophy. In this philosophy particular propositions are inferred from the phenomena and afterwards rendered general by induction.' (Cajori's translation; my italics).[27]

As this text shows, Newton was 'certain' that gravity 'penetrates to the very centre' of matter. He also knew however that assuming an attracting force inherent in matter itself would amount to admitting action at a distance. Physical events would influence one another without there being any physical link between them. He considered that so absurd that in his famous letter to his friend Bentley of January 17, 1692 he repudiated it in strongest terms as an interpretation of his own theory. He wrote there:

---

27 Cajori, *Sir Isaac Newton's Mathematical Principles of Natural Philosophy and his System of the World*, vol. II, p. 546 f. - Before Newton, Descartes had already rejected the idea of gravity being a force innate in matter, and tried to explain it through an ether hypothesis assuming three substances consisting of round particles with different sizes. See *Œuvres de Descartes*, vol. VIII, p. 202ff.; vol. IX, pp. 201ff.; also in vol. II, his correspondence with Mersenne, letters of June 19, 1639, par. 9 (p. 565) and August 27, 1639 (p. 569–573).

'It is inconceivable, that inanimate brute matter, should, without the mediation of something else, which is not material, operate upon and affect other matter without mutual contact, as it must be, if gravitation, in the sense of Epicurus, be essential and inherent in it. And this is the reason why I desired you would not ascribe innate gravity to me. That gravity should be innate, inherent, and essential to matter, so that one body may act upon another at a distance through a vacuum, without the mediation of any thing else, by and through their action and force Which may be conveyed from one to another, to me is great so to absurdity, that I believe no man, who has in philosophical matters a competent faculty of thinking, can ever fall into it. Gravity must be caused by an agent acting constantly according to certain laws, but whether this agent be material or immaterial, I have left to the consideration of my readers.' [28]

In the main text of his *Principia*, however, Newton did not mention that he had left gravity unexplained, but assumed it to be a force of attraction[29].

Others did not overlook the problem nevertheless. Goethe saw it and let Faust, the hero of his tragedy, yearn to 'detect the inmost force which binds the world and guides its course'[30]. Michael Faraday considered the hypothesis of an attracting force of matter unsatisfactory, but admitted being unable to find a better one[31]. No later scientist revealed the cause of that force. Nor did anyone succeed in finding a common principle explaining the external force of gravity and the internal forces of the atom, holding together the nucleus and determining the orbits of electrons. Some surmise that there must be a special particle acting as the carrier of gravity.

---

28 Quoted from Cajori, *Sir Isaac Newton's Mathematical Principles of Natural Philosophy and his System of the World*, vol. II, p. 633 f.

29 Cajori, *Sir Isaac Newton's Mathematical Principles of Natural Philosophy and his System of the World*, vol. II, Propositions I ff.

30 *Faust A Tragedy*, Bayard Taylor's translation (1870/1871).

31 See Faraday's letter to Reverend Jones in Peter Day, *The Philosopher's Tree, A Selection of Michael Faraday's Writings*, p. 104.

Decades ago they already gave it the name 'graviton'[32], yet it remained undiscovered. Indeed, in its present state theoretical physics is further removed than ever from solving the problem of gravitation. The family of particles of which the nucleus is believed to be composed has grown larger and larger. Some of them have even been split in smaller 'quarks' which again have different states or properties attributed to them. But none of them will open anything like an approach to explaining gravity.

Seen from the angle of the theory of science, the fact that some theoretical physicists nevertheless consider Einstein's General Theory of Relativity to have overcome Newtonian physics is more bewildering even than the fact that those questions remain unanswered after so many centuries. Events of that sort are highly unusual. I could name no parallel comparing to this in the history of science. Usually scientists will reject a theory that stood the test only if they have a better one replacing it. But many theoretical physicists consider Einstein's general theory of relativity to have overcome Newtonian theory although there never was a theory describing physical processes more precisely than Newton's does, and although they recognize that the general theory does *not* explain gravity. How could such a situation arise? We are facing a problem of first order in the history of thought. A greater challenge for the theory of science is hardly conceivable[33].

II

Where are we going astray? Why did we make no headway in more than three centuries? We might find an approach to that question by trying to understand why Einstein failed.

---

32 Stephen Hawking (*A Brief History of Time*, p. 157 ff,) quite naturally presupposes there that ,gravitons' must be the carriers of gravity.

33 In ,*Popper* versus *Einstein*' (ch. 7, p. 141ff.), I proposed a solution for the problem of gravity, which I will discuss in *Chapter 11*.

Some of his biographers tried to explain that failure by a change in Einstein's general outlook. Albrecht Fölsing assumed that observation and experiments had been his main concern as a young scientist, but that he he increasingly turned to purely mathematical principles in later years[34]. In that context he quoted from Einstein's Herbert Spencer Lecture *On the Method of Theoretical Physics*[35], given at Oxford in 1933, a passage he considered symptomatic.

Einstein said there 'that the axiomatic basis of physics cannot be an inference from experience, but must be free invention'. And he expressed his belief that 'pure mathematical construction enables us to discover the concepts and the laws connecting them', and that 'pure thought is competent to comprehend the real''[36].

Fölsing, himself a qualified physicist and usually reporting on the theory of relativity with unrestrained admiration, for once was critical of Einstein's view at this point and saw in it a 'vast over-estimation of pure thought in matters of the knowledge of nature.'[37] For the years after 1933 he painted the image of a man of considerably impaired judgment, appearing 'outside his field, weak, indecisive and contradictory', and taking little notice of the great discoveries of his time even within his field, being convinced that 'they do not seem … , at present, to make understanding the foundation any easier.'[38]

Could this interpretation be correct? At the time in question Einstein had not yet reached the age of 55 years. He was in good physical health, drank no alcohol, was on no other drugs and neither then nor later showed any symptoms of a mental disease. Going by medical experience he should

---

34 Albrecht Fölsing, *Albert Einstein - Eine Biographie*, pp. 761, 779, 780.

35 Einstein, *On the Method of Theoretical Physics*, Philosophy of Science, vol. 1 (1934), pp. 163–169.

36 Einstein, *On the Method of Theoretical Physics*, Philosophy of Science, vol. 1 (1934), p. 167; Fölsing *Albert Einstein - Eine Biographie*, p. 759–761, 779, 780.

37 Fölsing *Albert Einstein - Eine Biographie*, p. 759.

38 My translation from Fölsing, *Albert Einstein - Eine Biographie*, p. 780. Einstein presumably was referring to Edwin Hubble's discovery that the redshift always is in the direction of lower frequencies (1929), but the temporal sequence of events is not clear.

have been at the height of his mental abilities or at least at that of his discernment. What might explain such an early decline of his mental faculties then? Did it really exist at the time, or could it be that not his abilities had declined but his estimation in the world of science? Moreover, even if that were so, then what about his earlier performances, especially his theories of relativity? Were they really the products of a genius? Was Einstein a genius only temporarily then? Or had his theories also been conceived by a man of somewhat limited power of judgement?

At first glance those questions seem to be interesting only for psychologists and historians of science, but they also raise questions concerning the theory of science. How can it be that most physicists take Einstein's theory of relativity to be correct, even unchallengeable and exempt from all kinds of criticism, while scientists from other disciplines have greatest difficulties in understanding even the special theory, to say nothing of the far more complicated general theory? Are the brains of physicists structured differently from those of common people? And if so, then whose opinion is to count? Is physics only for physicists or it is everyone's concern? Is it possible that only young people can learn to understand the theory of relativity?

The first decades of the 20ᵗʰ Century were no good period for impartially discussing such issues. Initially, anti-Semitism was less of a hindrance than the strained relationships between Germany and her opponents of World War I. Who in Germany, on the eve of war with France, would have been willing to accept a criticism brought forward in 1913 by the French physicist Georges de Sagnac against the achievements of a scientist whom Germany claimed as her own?[39] And who would have criticized that scientist after 1918, when on him rested the hope of German science to find a way out from the isolation caused by the war, and back to international repute? Who would have offered opposition when this 'German' scientist, carrying a Swiss passport, became world-famous through

---

39 Georges de Sagnac, *L'éther lumineux, démontré par l'e effet du vent relatif d'éther* (Comptes rendus vol. 157 [1913[], 708) and *Sur la preuve de la réalité de l'éther lumineux* (Comptes rendus vol. 157 [1913], 1410). Sagnac there proposed an experiment permitting a very simple empirical test of Einstein's hypothesis of the constancy of the speed of light. I will discuss it in more detail in *Chapter 5, II, 4.*

his recognition by the British Royal Society after Eddington's experiment of 1919? Of all the representatives of German science, Einstein was the most important at that period. In Germany he was considered a 'Kultur-faktor'; his personal affairs had become a matter of concern for the government. And the converse situation must have existed in Great Britain and the United States. Criticizing Einstein then might have seemed like kicking the Germans when they were down. And in later years, criticizing a Jewish scientist might have looked like encouraging oncoming National Socialism in Germany.

All this was understandable in the historical situation but it was not conducive to sober scientific discussion. More than sixty years after Einstein's death and more than a century after his first publication of the special theory of relativity, time should have come for openly speaking one's mind on all those questions, and for discussing impartially Einstein's theories and the assumptions from which they started. Time should also have come for learning from Immanuel Kant's *Critique of Pure Reason* and Karl Popper's *Logic of Scientific Discovery* that even theoretical physicists must begin to think about the limits of human reasoning.

The real problems of theoretical physics were not at the level of mathematics; they consisted in understanding its limitations. The abilities of theoretical physicists in dealing with mathematical formulae were beyond doubt. In early 20th Century however, some of the most famous and influential among them wanted to be philosophers at the same time, though without having to learn about philosophy[40]. They did not try to

---

40 My criticism following in the text particularly refers to Bohr and Heisenberg, but we will see in *Chapters 3, III* and *9, I, 3* that Planck would also have deserved it. As to Bohr and Heisenberg: In Bohr's *Atomic Theory and Description of Nature*, vol. I, and in his papers *Can Quantum-Mechanical Description of Physical Reality be Considered Complete?*, Phys. Rev. 48 (1935), pp. 696 ff., and *On the Notions of Causality and Complementarity*, Dialectica 1948, pp. 312 ff., he proposed to introduce Hegelian dialectics into physics, and to bridge logical contradictions by assuming a 'dualism' of phenomena. Heisenberg accepted that approach for instance in *Physic and Philosophy* (1958). In Chapter III, dealing with the Copenhagen interpretation of quantum theory, he wrote there: ‚The dualism of two different descriptions of reality quite generally can be considered a fundamental difficulty no longer since *we know from the mathematical statement of the theory that it can contain no contradictions*' (p. 33 of the German translation; my re-translation and italics).

benefit by the results achieved in other fields of knowledge, particularly in the field of the theory of science. Kant would occasionally be mentioned, but his critically rational approach made itself felt nowhere in their own approaches to the problems of theoretical physics. Some of the greatest scientists and philosophers of the past centuries they largely ignored, Poincaré, Whitehead, Russell, Tarski, Gödel and Popper among them. Thus they did not benefit by the mistakes which others had made before them, but repeated them uncritically by making their physical theories depend on whether philosophical issues were to decide this way or that way, yet without discussing those issues. As a result, they ultimately sacrificed even mathematics and logic on the altar of their love for philosophy. That was a fatal mistake because scientists from other disciplines also deserve being taken seriously.

Einstein himself may have deserved that reproach less than others. He made considerable efforts to incorporate the philosophical views of his time in his own thinking, pondering deeply on epistemology and publishing his views in various papers. Those papers may help us to a better understanding not only of his own work as a physicist but also of the reactions of his contemporaries to the new theories he proposed.

CHAPTER 2:

A THIRD WAY TO COGNITION

Basic questions of epistemology. I. Deduction vs. induction. II. Theory of induction refuted; Popper: theory prior to observation. III. Einstein's axiomatic approach in Geometry and Experience *and* On the Method of Theoretical Physics; *more questions than answers.*

Discussing Einstein's views on the theory of science is no easy task because it will be necessary to make visible the mistakes they contain, and to demonstrate their influence on the theories he proposed. This chapter will show that inconsistency was his greatest problem in that field. His philosophical writings take up many ideas held modern in his time, and they thus hold attractions for anyone inclining towards those views. But to almost any interpretation there will also be passages proving the contrary, often in the same paper. That is why it never is convincing to pick on isolated statements confirming or disproving Einstein. A mere exegesis of his texts will hardly lead anywhere. For critical discussion it will be important, rather, to show the conflicting undercurrents in his thoughts. Even for that however, we must begin by looking at the background of the problems he was trying to solve.

I

What do we know and how certain is our knowledge? What knowledge can human beings own at all? What is the specific feature of *scientific* knowledge now accompanying technological progress and facilitating human life for so long?

Questions of that type were fundamental to the problems of epistemology for a long time, some of them since antiquity, and they remain unchanged in principle to this day. We can safely assume those questions, or at least something fairly close to them, to have been uppermost also in Einstein's mind when he embarked on his philosophical endeavours in early 20th Century.

Before going into details of his own views, we must try to trace back the ideas from which they originated. Various different approaches have been taken to the problems of epistemology, but only two of them became classical in the history of philosophy, the *deductive* approach and the *inductive* approach. In a sketchy typecast they can be described like this.

(*1*) A *deductivist* will assume that we gain our knowledge of nature by logical inference from major premises which we found to be true, and which we apply to specific situations. Following that notion, the 'scientific' knowledge of a natural scientist would consist in knowing those major premises. Newton's second law of motion, stating that

> 'the change of motion is proportional to the motive force impressed; and is made in the direction of the right line in which that force is impressed',

may serve as an example. It is a typical statement expressing theoretical physical knowledge. In the deductivist view, applying that knowledge consists in allocating to that major premise a description of facts in the shape of minor premises which will then by way of inference yield the assessment of the specific situation. In our example the minor premises would contain statements on specific operands, such as the performance of an engine and the weight of a vehicle. Combining with Newton's second law the values of those variables for 'motion' and for 'force' would then permit to calculate the 'change of motion' (acceleration or deceleration) of some specific vehicle.

The deductivist can claim that the rules of inference employed in this procedure satisfy every requirement of strictest formal logic. But the question of how he had reached his major premise and established its truth will embarrass him. It never has been possible to prove by purely logical means the truth of major premises such as Newton's second law of motion, or of other statements of similar type. For establishing the truth of his major premises the deductivist must fall back on other reasons.

What might those reasons be? Referring merely to the truth of the premise being 'evident' will hardly convince anyone where complex issues are at stake. If the deductivist instead refers to intellectual superiority which he claims for himself, or to other authorities such as his scientific education, or personal intuition, or divine revelation, then rational discussion would end at that point. Others too, at least theoretically, might have had scientific training, or been endowed with intuition, or been chosen for revelation. Clearly, arguments of that type provide no criteria for deciding between conflicting major premises.

A deductivist will need a criterion for distinguishing between conflicting insights, or revelations, or intuitions, and his theory does not yield that criterion. The answer to the question of the truth of his major premises is not part of his theory as it were. The possibilities of controlling by means of pure logic the truth of inferences drawn from such major premises also come to an early end. Everything will depend on the accuracy of the major premise which the deductivist cannot derive from an established truth. From the point of view of epistemology the question even seems to arise whether the deductive approach will gain us anything at all.

(2) The *inductivist*, by contrast, has long seen all that, and he thinks he knows better. Since the deductive approach cannot yield *certain* knowledge, his inference is, razor-sharp, that everything must in fact be quite different. Being usually convinced of Newton's infallibility, he will claim that

'whatever is not deduced from the phenomena is to be called an hypothesis; and hypotheses, whether metaphysical or physical, whether of occult qualities or mechanical, have no place in experimental philosophy. *In this philosophy particular propositions are inferred from the phenomena and afterwards rendered general by induction.*[41] (My italics)

According to that view, human knowledge does not consist in the abstract knowledge of universal laws but in experience gained from observation, or in theories inferred from experience.

A genuine inductivist will assert that in reality human knowledge does not start from abstract major premises but from specific observations, and that by generalizing their essential features we infer from them the underlying principles, and that these principles will then in turn enable us to predict future events to be expected under similar conditions. Repetition is an important feature of inductivist theory. The inductivist will claim that uniform events provide the material from which we infer the laws of nature. For gaining knowledge reliably secured and meriting the attribute 'scientific', the inductivist scientist will strive at obtaining as large as possible a number of carefully made observations. The broader and securer his empirical basis, so he will claim, the better reliable will be the conclusions drawn from it.

(3) That approach, however, will get the inductivist also in logical difficulties which David Hume (1711–1776) already demonstrated in principle[42]. By the rules of logic no statement based on experience can yield conclusions going beyond that experience. If the inductivist observed in a thousand cases that Newton's second law of motion always was confirmed, then it still does not follow by valid inference that the same must happen also in the thousand and first case. His major premise, derived from observation, rests on a thousand cases but unfortunately not on thousand and one. For *proving* the general principle which he wants to derive by induc-

---

41 Cajori, *Sir Isaac Newton's Mathematical Principles of Natural Philosophy and his System of the World*, vol. II, p. 546 f. – I quoted the same passage more fully in *Chapter 1, I.*
42 David Hume, *A Treatise of Human Nature*, vol. I, part III, section VI.

tion from experience, the inductivist would have to claim that inductive inferences drawn from large numbers of empirical statements can yield abstract principles such as Newton's second law of motion. He would have to show a way of deriving from observations made in the past statements of events that are to happen in future. The general principle which he wants to infer from his observations is to apply not only in the Arabian Nights, and not only in millions or billions of cases, but forever, or at least to the end of this world. It will tolerate no exception. As long as we have no reliable knowledge of when the world will end, the size of the class to which that principle applies must therefore be *infinite*.

That obviously reaches far beyond the possible scope of human knowledge. But the logical problems of the inductivist will in fact be even greater than that, as Karl Popper showed in his *Logik der Forschung* (1934)[43]. By the rules of logic, ever so many observations confirming some law of nature will not even permit to state a *probability* of that law being true; not, at any rate, if by 'probability' we mean the mathematical ratio of the number of cases already observed to those still to happen in future. If mathematical ratios are to be relevant for expressing probabilities, then the probability of any statement of a law of nature being true must always be zero because the ratio to infinity of any number of cases is always zero, no matter how great that number may be. Mathematically the equation $1 : \infty = 0$ is as correct as the equation $10^{50} : \infty = 0$. Not even trillions of reliable observations, indeed not even all observations man ever has made or will make, can alter the fact that the mathematical probability determined by the ratio of the number of observations to that of expectations must always be zero. That simply follows from the fact that we are unable to foresee the future. As long as that remains so, every law of nature which we believe to have discovered can only be in the nature of a *hypothesis* as Popper convincingly showed.

---

43 Karl Popper, *The Logic of Scientific Discovery*, there in particular *Two Notes on Induction and Demarcation*, p. 311 ff.; also *Conjectures and Refutations*, vol. I, p. 33 ff., 42 ff.

II

The conflict between deductivism and inductivism leads to the fundamental problem of epistemology.

Inductivism prevailed for a long time among natural scientists. The
statement quoted above from Newton's *Principia* shows that. In fact, going not by the strength of its arguments but by the number of its followers,
it probably still prevails in our time. But that only shows that insights
achieved in the theory of science will often remain unheeded in other
disciplines. As a possible approach to the theory of scientific knowledge
inductivism has been refuted in principle since Kant's *Kritik der reinen
Vernunft* (1781) and in detail since Popper's *Logik der Forschung* (1934).

It is incompatible with logic, and contrary to reality. Its basic allegation that we *in fact* gain our knowledge of the laws of nature by way of
induction from experience simply is not true, even if Newton claimed it
to be so. He did not see that the very observation leading to experience is
possible only *after* we have chosen a point of view from which to observe.
Observation can only be in the light of an aspect we choose. And where
laws of nature are concerned we never start from observation in reality,
but always from a point of view supplied by a theory. As Karl Popper once
put it,

'without waiting for premises we jump to conclusions,'[44]

His explanation of that relationship is vivid and clear. Counter-arguments
deserving consideration are unknown to this day. I will not repeat it in
more detail here because we will often be back with it in this book, and
because Popper's works deserve being read in the original. A short passage
from a paper he read at Cambridge in 1953 will be sufficient for illustrating the point. He said there:

---

44 Karl Popper, *Conjectures and Refutations*, p. 46.

'The belief that science proceeds from observation to theory is still so widely and so firmly held that my denial of it is often met with incredulity. I have even been suspected of being insincere – of denying what nobody in his senses can doubt.

But in fact the belief that we can start with pure observations alone, without anything in the nature of a theory, is absurd; as may be illustrated by the story of the man who dedicated his life to natural science, wrote down everything he could observe, and bequeathed his priceless collection of observations to the Royal Society to be used as inductive evidence. This story should show us that though beetles may profitably be collected, observations may not.

Twenty-five years ago I tried to bring home the same point to a group of physics students in Vienna by beginning my lecture with the following instructions: "Take pencil and paper; carefully observe, and write down what you have observed!". They asked, of course, what I wanted them to observe. Clearly the instruction "Observe!" is absurd. (It is not even idiomatic unless the object of the transitive verb is taken as understood.) Observation is always selective. It needs a chosen object, a definite task, an interest, a point of view, a problem.' [45]

That simple argument convincingly and finally refutes the inductivist assertion that the process of scientific understanding of nature consists *in fact* in inferring the laws of nature from experience. There never has been or will be a valid theory of knowledge based on induction. That will be my premise here. Readers not accepting it I must refer to Popper's own works, especially to his *Logic of Scientific Discovery* and to the majority of his essays in *Conjectures and Refutations*.

In Einstein's time however, and due to Newton's overwhelming influence in physics, inductivism still was widely accepted among scientists despite the questions it leaves open. In 1895 the University of Vienna even established a special chair for 'Philosophy, in Particular the History and Theory of the Inductive Sciences', Ernst Mach being the first to hold

---

45 Karl Popper, Models, *Instruments and Truth* in *The Myth of the Framework*, p. 154, 155.

it[46]. As the name shows, one took for granted then that the natural sciences must be inductive sciences. Some members of the 'Vienna Circle' believed inductivism to be correct so doubtlessly that they no longer saw the problem of epistemology in the 'if' but only in the 'how' of inductive reasoning, and were therefore primarily concerned with trying to establish inductive inference as a special type of logical inference. Even in our days many scientists consider inductivism the natural and only correct theory of scientific knowledge.

III

We have now seen in broadest outline the background of ideas before which Einstein developed his theory of scientific knowledge. A deep-rooted belief in inductivism was one of its elements. His own views do not readily fit that pattern however. He was neither deductivist nor pure inductivist. Deduction he not even mentioned as a possible source of scientific knowledge in his essays. Induction he seems to have considered at least a possible way of gaining it. Nevertheless it was not his way.

Einstein believed, rather, that he had found a third way of solving the major problems of epistemology at least for the field of physics, a way he believed to be fundamentally different from deductivism and inductivism and better than both of them. His royal road to cognition consisted in applying to physics the *axiomatic method of geometry*. That method, to which geometry not only owed its amazing success but also its unmatched power of persuasion, would be the key also to physical knowledge, so he must have thought. Transferring to physics that method, and strictly observing its rules would then secure physical science the rank of being an absolutely 'exact science'.

In this section I will try to state Einstein's epistemological approach in a way that will make it intelligible and may later serve as a footing for

---

46   Dieter Hoffmann, *Ernst Mach* in v. Meÿenn, *Die Großen Physiker*, vol. II, p. 26.

criticism. Einstein dealt with questions of epistemology in several lectures and papers. Apart from differences in focus and wording the views he put forward remained the same in principle[47]. For illustrating them it will therefore be sufficient to quote from only two papers which are his most important in that field. The quotations will have to be lengthy however, in order to show also the context of his thoughts.

Understanding Einstein's philosophical texts is not easy. I contend that their difficulty mainly results from the fact that they are contradictory in themselves. I will comment only briefly on those texts in this chapter, and discuss them in more detail in those following. Based on the passages quoted below critical readers will thus be able to shape their own opinion on Einstein's theory of knowledge. I hope they will notice that the origin of those contradictions is not in my interpretation but in his own words.

(*1*) Speaking to the Prussian Academy of Sciences at Berlin in 1921, Einstein read a paper on *Geometry and Experience*. He probably framed his epistemological convictions for the first time in a broader context there. To facilitate discussion I have put in italics the passages that will later get special mention[48].

> 'One reason why mathematics enjoys special esteem, above all other sciences, is that its laws are *absolutely certain and indisputable*, while those of all other sciences are to some extent debatable and in constant danger of being overthrown by newly discovered facts. In spite of this, the investigator in another department of science would not need to envy the mathematician if the laws of mathematics referred to objects of our mere imagination, and not to objects of reality. For it cannot occasion surprise that different persons should arrive at the same logical conclusions when they have already agreed upon the fundamental laws (axioms), as well as the methods by which other laws are to be deduced therefrom.

---

47 The most important papers are now in *The Collected papers of Albert Einstein*.

48 Albert Einstein, *Geometrie und Erfahrung* (Geometry and Experience), Sitzungsberichte der Preußischen Akademie der Wissenschaften, 1921, p. 882–883 (translation quoted from *Collected papers of Albert Einstein*).

But there is another reason for the high repute of mathematics, in that it is mathematics which affords the exact natural sciences a certain measure of security, to which without mathematics they could not attain.

At this point an enigma presents itself which in all ages has agitated inquiring minds. *How can it be that mathematics, being after all a product of human thought which is independent of experience, is so admirably appropriate to the objects of reality?* Is human reason, then, without experience, merely by taking thought, able to fathom the properties of real things?

In my opinion the answer to this question is, briefly, this: *As far as the laws of mathematics refer to reality, they are not certain; and as far as they are certain, they do not refer to reality.* It seems to me that complete clearness as to this state of things first became common property through that new departure in mathematics which is known by the name of mathematical logic or 'axiomatics'. The progress achieved by axiomatics consists in its having neatly separated the logical formal from its objective or intuitive content; according to axiomatics the logical-formal alone forms the subject-matter of mathematics, which is not concerned with the intuitive or other content associated with the logical-formal.

Let us for a moment consider from this point of view any axiom of geometry, for instance, the following: Through two points in space there always passes one and only one straight line. How is this axiom to be interpreted in the older sense and in the more modern sense?

The older interpretation: Every one knows what a straight line is, and what a point is. Whether this knowledge springs from an ability of the human mind or from experience, from some collaboration of the two or from some other source, is not for the mathematician to decide. He leaves the question to the philosopher. Being based upon this knowledge, which precedes all mathematics, the axiom stated above is, like all other axioms, self-evident, that is, it is the expression of a part of this *a priori* knowledge.

The more modern interpretation: Geometry treats of entities which are denoted by the words straight line, point, etc. These entities do not take for granted any knowledge or intuition whatever, but they presuppose only the validity of the axioms, such as the one stated above, which are to be taken in a purely formal sense, i.e. as void of all content of intuition or experience. *These axioms are free creations of the human mind.* All other propositions of geometry are logical inferences from the axioms (which are to be taken in the nominalistic sense only). The matter of which geometry treats is first defined by the axioms. Schlick in his book on epistemology has therefore characterised axioms very aptly as "implicit definitions".

This view of axioms, advocated by modern axiomatics, purges mathematics of all extraneous elements, and thus dispels the mystic —obscurity which formerly surrounded the principles of mathematics. *But a presentation of its principles thus clarified makes it also evident that mathematics as such cannot predicate anything about perceptual objects or real objects.* In axiomatic geometry the words "point", "straight line", etc., stand only for empty conceptual schemata. That which gives them substance is not relevant to mathematics.

Yet on the other hand it is certain that mathematics generally, and particularly geometry, owes its existence to the need which was felt of learning something about the relations of real things to one another. The very word geometry, which, of course, means earth-measuring, proves this. For earth-measuring has to do with the possibilities of the disposition of certain natural objects with respect to one another, namely with parts of the earth, measuring-lines measuring-wands, etc. It is clear that the system of concepts of axiomatic geometry alone cannot make any assertions as to the relations of real objects of this kind, which we will call practically-rigid bodies. *To be able to make such assertions, geometry must be stripped of its merely logical-formal character by the co-ordination of real objects of experience with the empty conceptual framework of axiomatic geometry. To accomplish this, we need only add the proposition:*

Solid bodies are related, with respect to their possible dispositions, as are bodies in Euclidean geometry of three dimensions. Then the propositions of Euclid contain affirmations as to the relations of practically-rigid bodies.

Geometry thus completed is evidently a natural science; we may in fact regard it as the most ancient branch of physics. *Its affirmations rest essentially on induction from experience, but not on logical inferences only.* We will call this completed geometry "practical geometry," and shall distinguish it in what follows from "purely axiomatic geometry." The question whether the practical geometry of the universe is Euclidean or not has a clear meaning, and its answer can only be furnished by experience. All linear measurement in physics is practical geometry in this sense, so too is geodetic and astronomical linear measurement, if we call to our help the law of experience that light is propagated in a straight line, and indeed in a straight line in the sense of practical geometry.'

This text shows that Einstein wanted to transfer the axiomatic method of Euclidean geometry to physics directly and unaltered. Adding to it one more axiom was to convert the non-empirical discipline of geometry into the empirical science of physics. And the new axiom was to consist in the statement that 'solid bodies are related, with respect to their possible dispositions, as are bodies in Euclidean geometry of three dimensions'.

The text shows less clearly how we could obtain the empirical knowledge of laws of nature as taught by physical science. Einstein said of those laws that their 'affirmations rest essentially on induction from experience, but not on logical inferences only'. That seems to suggest that he considered induction from experience a possible way of logical inference. He told us no more about that however. He did not say how to overcome Hume's argument that induction from statements on experience will not carry inferences going beyond that experience. Instead, he only referred to 'practical geometry' in the end, but without saying what kind of mental operation might permit inferring from the limited experience, which human beings only can have, the knowledge of natural laws that are to be valid indefinitely. Nor did he answer that question in the further text of that paper.

(2) In a later paper Einstein treated the same subject in more detail. His already mentioned Herbert Spencer Lecture, read at Oxford in 1933, was

On the Methodology of Theoretical Physics. In the following quotation I have again put in italics passages that will later get special mention[49].

'I want now to glance for a moment at the development of the theoretical method, and while doing so especially to observe the relation of pure theory to the totality of the data of experience. Here is the eternal antithesis of the two inseparable constituents of human knowledge, Experience and Reason, within the sphere of physics.

We honour ancient Greece as the cradle of western science. She for the first time created the intellectual miracle of a logical system, the assertions of which followed one from another with such rigor that not one of the demonstrated propositions admitted of the slightest doubt – Euclid's geometry. This marvellous accomplishment of reason gave to the human spirit the confidence it needed for its future achievements. *The man who was not enthralled in youth by this work was not born to be a scientific theorist.*

But yet the time was not ripe for a science that could comprehend reality, was not ripe until a second elementary truth had been realised, which only became the common property of philosophers after Kepler and Galileo. *Pure logical thinking can give us no knowledge whatsoever of the world of experience; all knowledge about reality begins with experience and terminates in it.* Conclusions obtained by purely rational processes are, so far as Reality is concerned, entirely empty. It was because he recognized this, and especially because he impressed it upon the scientific world that Galileo became the father of modern physics and in fact of the whole of modern natural science.

But if experience is the beginning and end of all our knowledge about reality, what role is there left for reason in science?

A complete *system of theoretical physics* consists of concepts and basic laws to interrelate those concepts and of consequences to be derived by logical deduction. It is these consequences to which our particular experiences are to correspond, and it is the logical

49 Einstein, *On the Methodology of Theoretical Physics*, Philosophy of Science, Vol. 1, No. 2 (1934), pp. 163–169.

derivation of them which in a purely theoretical work occupies by far the greater part of the book.

This is really exactly analogous to Euclidean geometry, except that in the latter the basic laws are called 'axioms'; and, further, that in this field there is no question of the consequences having to correspond with any experiences. *But if we conceive Euclidean geometry as the science of the possibilities of the relative placing of actual rigid bodies and accordingly interpret it as a physical science, and do not abstract from its original empirical content, the logical parallelism of geometry and theoretical physics is complete.*

We have now assigned to reason and experience their place within the system of theoretical physics. Reason gives the structure to the system; the data of experience and their mutual relations are to correspond exactly to consequences in the theory. *On the possibility alone of such a correspondence rests the value and the justification of the whole system, and especially of its fundamental concepts and basic laws. But for this, these latter would simply be free inventions of the human mind which admit of no a priori justification either through the nature of the human mind or in any other way at all.*

The basic concepts and laws which are not logically further reducible constitute the indispensable and not rationally deducible part of the theory. *It can scarcely be denied that the supreme goal of all theory is to make the irreducible basic elements as simple and as few as possible without having to surrender the adequate representation of a single datum of experience.*

The conception here outlined of the purely fictitious character of the basic principles of theory was in the eighteenth and nineteenth centuries still far from being the prevailing one. But it continues to gain more and more ground because of the ever-widening logical gap between the *basic concepts and laws* on the one side and the consequences to be correlated with our experiences on the other – a gap which widens progressively with the developing unification of the logical structure, that is with the reduction in the number of the logically independent conceptual elements required for the basis of the whole system.

Newton, the first creator of a comprehensive and workable system of theoretical physics, still believed that the basic concepts and laws of his system could be derived from experience; his phrase '*hypotheses non fingo*' can only be interpreted in this sense.

In fact at that time it seemed that there was no problematical element in the concepts, Space and Time. The concepts of mass, acceleration, and force and the laws connecting them, appeared to be directly borrowed from experience. But if this basis is assumed, the expression for the force of gravity seems to be derivable from experience; and the same derivability was to be anticipated for the other forces.

One can see from the way he formulated his views that Newton felt by no means comfortable about the concept of absolute space, which embodied that of absolute rest; for he was alive to the fact that nothing in experience seemed to correspond to this latter concept. He also felt uneasy about the introduction of action at a distance. But the enormous practical success of his theory may well have prevented him and the physicists of the eighteenth and nineteenth centuries from recognizing the fictitious character of the principles of his system.

On the contrary the scientists of those times were for the most part convinced that the basic concepts and laws of physics were not in a logical sense free inventions of the human mind, but rather that they were derivable by abstraction, i.e. by a logical process, from experiments. It was the general Theory of Relativity which showed in a convincing manner the incorrectness of this view. For this theory revealed that it was possible for us, using basic principles very far removed from those of Newton, to do justice to the entire range of the data of experience in a manner even more complete and satisfactory than was possible with Newton's principles. But quite apart from the question of comparative merits, the fictitious character of the principles is made quite obvious by the fact that it is possible to exhibit two essentially different bases, each of which in its consequences leads to a large measure of agreement with experience. This indicates that any attempt logically to derive the basic concepts and laws of mechanics from the ultimate data of experience is doomed to failure.

If then it is the case that the axiomatic basis of theoretical physics cannot be an inference from experience, but must be free invention, have we any right to hope that we shall find the correct way? Still more-does this correct approach exist at all, save in our imagination? Have we any right to hope that experience will guide us aright, when there are theories (like classical mechanics) which agree with experience to a very great extent, even without comprehending the subject in its depths? To this I answer with complete assurance, that in my opinion there is the correct path and, moreover, that it is in our power to find it. *Our experience up to date justifies us in feeling sure that in Nature is actualized the ideal of mathematical simplicity. It is my conviction that pure mathematical construction enables us to discover the concepts and the laws connecting them which give us the key to the understanding of the phenomena of Nature.* Experience can of course guide us in our choice of serviceable mathematical concepts; it cannot possibly be the source from which they are derived; *experience of course remains the sole criterion of the serviceability of a mathematical construction for physics, but the truly creative principle resides in mathematics. In a certain sense, therefore, I hold it to be true that pure thought is competent to comprehend the real, as the ancients dreamed.'*

Apart from elements already known from Einstein's lecture on '*Geometry and Experience*' this text shows some new and interesting ideas.

That particularly concerns the role assigned to experience. It seems to play a dual role in Einstein's theory, being both the starting point and the criterion of knowledge. He said 'In a certain sense, therefore, I hold it to be true that pure thought is competent to comprehend the real, as the ancients dreamed.' Yet he also said 'Pure logical thinking can give us no knowledge whatsoever of the world of experience; all knowledge about reality begins with experience and terminates in it.'

How are such views to be reconciled with one another? Is it possible that human knowledge could arise from pure thought, yet begin with experience and terminate in it? How can we gain such knowledge? In both papers Einstein explicitly refused to recognize any knowledge valid *a priori* [50]. He

---

50 In *The Problems of Space, Fields and Aether in Physics* Einstein expressed his repudiation of knowledge *a priori* even more clearly (Transactions of the 2nd World power

considered the 'fundamental concepts and basic laws' underlying the 'system of theoretical physics' as being 'free inventions of the human mind which admit of no *a priori* justification either through the nature of the human mind or in any other way.' Nevertheless, when claiming in the last paragraph of that quotation 'that pure mathematical construction enables us to discover the concepts and the laws connecting them which give us the key to the understanding of the phenomena of nature', he seemed to refer to the 'fundamental concepts and basic laws' which he had mentioned before. He seemed to arrogate for those basic laws the certainty of mathematics. But how can empirical knowledge be a 'free invention of the human mind' and at the same time be 'based on pure mathematical construction'? At first glance Einstein's texts thus leave open many questions.

conference, Berlin (1930), 19, 1–5.

CHAPTER 3:

THE POVERTY OF AXIOMATISM

Kant's sapere aude! I. Axioms in traditional understanding; Euclidean geometry the model of axiomatic understanding of science. II. Transfer of axiomatic method to empirical science problematic; logical inferences tautological, hence empirical discoveries through logic or mathematics impossible; Einstein's view of physics; his additional axiom; his view of the aims of science; practical application not his focus. III. Historical context of Einstein's view; dissemination of his understanding; origin of axiomatic understanding of science in Aristotelian tradition; Descartes' approach more geometrico; his influence on later centuries; distinction between logic, mathematics and empirical science; Whitehead-Russell's Principia Mathematica; Tarski's levels of language; conflict between non-Euclidean geometries and axiomatic approach to physics; primacy of mathematics, physics considered auxiliary science; conflict irresolvable but faith unbroken; relevance of Gödel's proof; incompatibility with theory of relativity unnoticed by Einstein.

Einstein's views on epistemology are not easy to understand. In fact, they quite often are rather confusing. Ostensively convincing at first glance, many of his considerations turn out to be dark and contradictory upon closer scrutiny. This chapter will show that his papers on the theory of science leave the important questions unanswered in the end.

That will become apparent only to those approaching Einstein's texts with the self-confidence needed for trusting their own judgement. A scientist modestly assuming that even if she or he cannot understand Einstein, others may yet be able to, takes the wrong attitude. Diffidence sometimes is hard to get over however, even in science. Reverence for great names will often be overwhelming, and that is a major problem in any critical discussion of Einstein's theories. Most of us were brought up to see in him a human miracle first, one of those almost supernatural beings whose achievements surpass anything mere mortals can ever hope to attain. Long before our own faculties of judgement even had a chance of recognizing impending questions as the problems they were, others had already imbued us with the notion that Albert Einstein was the supreme genius of the 20th Century, and that his theory of relativity was one of the greatest feats ever of the human mind. It is only natural if we tend to regard him with exaggerated respect, and to consider his thoughts the recipes of an outstanding authority in science.

Diffidence may be a symptom of modesty but it is no proper attitude for a scientist because it does not serve the progress of science[51]. In science there are experts but no authorities, as Karl Popper once said. Science requires intellectual independence, and intellectual independence requires irreverence, not to persons but to ideas. A scientist must have the courage to shape his opinions himself. He must take the risk of being mistaken and of exposing himself to criticism. That was why Kant framed the motto *sapere aude!* Dare to use your own brains! The bold and critical spirit of the Enlightenment for which he called must not stop at criticising Einstein's theories. We owe that to Einstein's memory if nothing else. He was no authority and did not claim to be one. His uncritical admirers only made him so. Like the adult spectators in Andersen's tale of *The Emperor's New Clothes* they found applauding him less perilous than trusting their own judgement. Einstein himself always was sincerely concerned about the progress of science. By his attack on Newtonian physics he made the greatest effort to overcome authoritarianism in science[52]. And he was not interested in styling himself a monument but was made so by circumstances, the ambition of the Prussian government probably being foremost

---

51 For Karl Popper's view on the importance of disagreement for the progress of science see, for instance, *The Myth of the Framework* p. x, 7, 16, 34, 91; also *The World of Parmenides*, p. 62 ff.

52 Karl Popper, *Science, Problems, Aims, Responsibilities*, in *The Myth of the Framework'*, p. 82 ff., 91; also *On the Sources of Knowledge and Ignorance*, sec. xv, in *Conjectures and Refutations*, p. 3 ff.

among them[53]. He was a professional scientist and had a high level of competence. But he, too, could be mistaken. And he knew that and admitted it whenever he noticed his own errors.

Anybody reading Einstein's texts critically and mustering the courage needed for independent thinking will see that they raise more questions than they answer. That need not imply, however, that we cannot derive their meaning from their context or from circumstances. The full import of an academic performance sometimes becomes visible only against the intellectual background of its origin. If we want to understand Einstein's theory of science, we must try to place ourselves in his world of ideas and to look at the problems he was trying to solve as he might have seen them in his time.

Einstein's notion of the *axiomatic method* and its possible applications must be an important key to understanding his view of the theory of science. His writings indicate that this method influenced his entire philosophical outlook, and his famous words 'God does not throw dice' show that it even affected his religious beliefs. It would be impossible to understand his attitude towards the theory of science without understanding that method.

Scrutiny will show that Einstein's understanding of the axiomatic method will ultimately not stand up to criticism. That does not mean, however, that his ideas did not include an important message which was right in principle and should be preserved and further developed if possible. Einstein wrestled for decades with the problem of scientific knowledge, and he pondered deeply over the axiomatic method. It would be presumptuous to assume that a man of his abilities had misunderstood just everything, and had contributed nothing at all to the solution of the problems that were on his mind.

My own view is, quite on the contrary, that Einstein made an extremely important contribution to modern theory of science. His heuristic approach which I will discuss in *Chapters 4* and *5* was revolutionary indeed. The actual problem seems to lie in the fact that this part of his thoughts,

---

53 Fölsing, *Albert Einstein - Eine Biographie*, 2nd. ed. (1993), p. 371–374.

the part where I think he was right, was so original and revolutionary that many scientists fail to understand it to this day[54]. The mistakes on the other hand, which he also made, came from a tradition of thought that was quite unoriginal but for that very reason speedily met with almost universal consent.

I

Before coming to that, we must look at the axiomatic method more closely.

Traditional understanding assumes axioms to be the basic truths of a science on which all other findings of that science rest. They are its 'fundamental propositions'. Although themselves not deduced from other propositions or proved in any other way, their truth is nevertheless considered to be certain. The explanation given for that certainty will mostly be that the truth of an axiom is 'directly apparent', needing no further justification therefore. Statements of Euclidean geometry, such as that 'through two points in space there passes one and only one straight line', will usually serve as examples[55]. We can leave it at that for the moment. Later we will see that axioms are in fact definitions, and that the standard for gauging them is not truth but expediency.

From sets of axioms we can derive other statements by logical inference. If a deduction is logically correct, then the conclusion will be as true or false as the premises from which it was drawn. If the premises of a deduction consist of axioms of which the truth is certain, or if they consist of statements deduced by correct inference from such axioms, then the truth of the conclusion thus drawn must be equally certain.

---

54 John Eccles was one of the rare exceptions; see his discussion of creative imagination in *Evolution Of The Brain Creation Of The Self*, ch. 10.5b.

55 Euclid referred to them as ‚postulates'.

All conclusions inferred from such axioms according to the rules of logic must be as true as the axioms themselves in this understanding. The consistency of those inferences as well as their mutual compatibility must also automatically arise from the fact that they follow from the same axioms, and that contradicting statements cannot be true at the same time.

Euclidean geometry with its infallible certainty of compelling evidence is the shining model of this axiomatic understanding of science. Its followers, one of whom Einstein avowed himself explicitly as we saw, strive to transfer to other fields of science the axiomatic method in order to obtain the same degree of certainty there[56]. Their notion is that assuming exact execution, and no formal error occurring in the process of deducing itself, this method must create, in any science whatsoever, a system of coherent and non-contradictory statements of which the truth depends only on that of the underlying axioms. Wittgenstein's statement that

> 'the totality of true propositions is the whole of natural science
> (or the totality of the natural sciences),'[57]

is characteristic of that understanding. Its followers believe logical contradictions in a fully axiomatized science to be *theoretically* impossible. For the field of Euclidean geometry David Hilbert could even prove that[58].

Seen from that angle and assuming axioms to be expressing empirical *truth*, error will not appear as a normal element in the evolution of science. Rather, it will seem a symptom of human weakness and personal deficiency, possibly even of dishonesty.

---

56 For Einstein's view see the text quoted in *Chapter 1, III, 1* from his *Geometry and Experience.*

57 Ludwig Wittgenstein, *Tractatus Logico-Philosphicus*, sec. 4.11.

58 Ernest Nagel/James R. Newman, *Gödels Proof*, p. 8 ff.

II

It is far from obvious that we could transfer this understanding of science, shaped from the model of axiomatic geometry, to empirical science as easily as Einstein thought. On the contrary, looking at it from the angle of our present knowledge, the approach meets with a serious objection raised by the groundbreaking works of Whitehead and Russell in their *Principia Mathematica* (1910). They seem to have been the first to realise that logical operations and mathematical equations are tautological in principle.

In mathematics the information must be the same on both sides of an equation. In logic, similarly, the information in the inference must also be in the premises from which it was drawn. Pure logic cannot extend the informational content of a system of propositions[59]. Logical operations can rearrange in different ways the information given in their premises, but they cannot yield information that was not contained in those premises. That is why computers can perform logical operations but make no empirical discoveries.

(1) If that is correct, then the obvious inference must be that for a science aiming at making empirical *discoveries* the axiomatic method cannot deserve discussing from the outset. In modification of Einstein's famous words in *Geometry and Experience* we can say that[60]

> if a statement can be deduced from premises previously known, then it contains no empirical discovery; and if it contains an empirical discovery, then it cannot have been deduced from premises previously known.

---

59 Alfred North Whitehead/Bertrand Russell, *Principia Mathematica*, particularly appendix C, *Truth-Functions and Others*.

60 *Chapter 1, III, 1.*

Anyone seeing that differently would first have to deal with Whitehead and Russell's theory of logic and with Karl Popper's *Logic of Scientific Discovery*.

Nevertheless the passages quoted in *Chapter 2, III* from Einstein's papers leave no doubt that his understanding of science was based on the axiomatic approach. Physics was to him, in Wittgenstein's words, a 'totality of true propositions'. He saw in it a collection of true statements, some known, others unknown and still to be discovered, but all of them initially true and free of contradictions. And he believed that all of them could be obtained by way of deduction from a small number of 'basic elements'. In his notion pure geometry was 'a product of human thought which is independent of experience' and limited to 'the logical-formal'. But it could 'be stripped of its merely logical-formal character by the co-ordination of real objects of experience with the empty conceptual frame-work of axiomatic geometry'. To that end, so he believed, we only had to increase the basic axioms of Euclidean geometry by one more axiom consisting in the proposition that

> 'solid bodies are related, with respect to their possible dispositions, as are bodies in Euclidean geometry of three dimensions.'

That extension by a single new axiom was to transform axiomatic geometry into the empirical science of physics [61].

(2) Einstein's papers show that his belief in the axiomatic foundation of science influenced not only his method of scientific research but even stronger the aims he pursued by it. In his view there was an important difference between the aims of experimental physics and those of theoretical physics as it had recently emancipated itself.

---

61 Einstein expressed this view also on other occasions, for instance in his paper *My Theory*, The Times, Nov. 13, 19190; also in *Das Raum-, Äther- und Feld-Problem der Physik* in *Mein Weltbild*, p. 229 ff., 233. - Popper considered possible Einstein's approach. See his *Logic of Scientific Discovery*, sec. 17. I will discuss that in more detail in *Chapter 4, III*.

An experimental physicist living in the 19th Century might have seen his task for example in investigating the phenomena of electricity. He might have conjectured that there may be mutual dependencies between electrical tension, current and resistance, and might have devised experiments for measuring them under different conditions. Thus he would ultimately have arrived at the law discovered by Georg Ohm in 1826, by which electrical current corresponds to the quotient of voltage and resistance ($I = V/R$).

Einstein always respected efforts of that kind. They were an essential part of physical research in his view. Yet they were not the ultimate goal of science to him, at least not of theoretical physics, which was not concerned with discovering new physical effects or with finding new applications for physical laws as he saw it, but with far more fundamental questions. Applying those laws was important of course, since it must show whether the principles developed by theoretical physics conformed to reality. For, as he said and believed, 'all knowledge about reality begins with experience and terminates in it'. He would tolerate no conflict between theory and experience. The more important task of theoretical physics, however, lay in revealing even deeper truths. It was to disclose the 'irreducible basic elements' of science, and to make them 'as simple and as few as possible', yet always 'without having to surrender the adequate representation of a single datum of experience'. The endeavours of the theoretical physicist were concerned with the system thus, but experience was to remain the supreme judge at all events.

The approach indicates that empirical discoveries were no goal of theoretical research in Einstein's understanding, but a criterion for assessing its results. They were no aim in themselves, but a means for achieving a broader goal. That broader goal consisted in discovering the common principles underlying experience and reducing them further and further. In another lecture he said

'The highest task of physicists is, then, the search for those most general elementary laws from which, by pure deduction, the world view can be gained. No logical path will lead to these elementary laws, but only intuition based on empathy in experience.'[62]

From those 'general elementary laws' of nature thus to be revealed everything else must then follow by itself in his view. For he was convinced that

'our experience up to date justifies us in feeling sure that in Nature is actualized the ideal of mathematical simplicity. It is my conviction that pure mathematical construction enables us to discover the concepts and the laws connecting them, which give us the key to the understanding of the phenomena of Nature. ... The truly creative principle resides in mathematics.'[63]

Going by his preceding explanations, the 'ideal of mathematical simplicity' invoked there could only consist in the axioms underlying the system.

My quotations ought to have shown that in Einstein's notion the whole of physics consisted in a complex system of true statements, some of them known already, others still to be discovered. Underlying them all, so he assumed, there was one common principle, unfortunately still largely unknown. We might try to visualize that notion by comparing the whole field of physics, known or unknown, with a very large and ornate carpet of which we only see some fragments at present, but without being able to recognize the pattern on which it was made. Had we once understood that pattern, then that would enable us to fill in the gaps still open. In the principle of the constancy of the speed of light and in the formula $e = mc^2$, which Einstein believed to be directly linked with it, the special theory of relativity had revealed a small element of that pattern[64]. By expanding it in the general theory to all types of motion including acceleration and curved lines, even more principles had come in sight. The truly universal

---

62 My translation from Albert Einstein *Prinzipien der Forschung*, Address at the occasion of Max Planck's 60th anniversary, in *Mein Weltbild*, p. 175 ff., 178.

63 Einstein, *On the Method of Theoretical Physics*, quoted in *Chapter 1, III, 2*.

64 For details, see *Chapter 9, VI*.

pattern however, underlying really everything, had not yet become visible.

The unified field theory for which Einstein searched in vain in the last decades of his life was then to have led the way to even more fundamental principles, ultimately even to the one universal formula, the 'world formula', now also called 'theory of everything', which he thought must be underlying all physical truths.[65] That would have been the discovery of the one fundamental axiom, 'the ideal of mathematical simplicity' actualized in nature. And once that had been found, the further completion of the axiomatic edifice of theoretical physics would have been as simple as completing a carpet on a known pattern. The benefit to science would be immense if instead of cautiously taking tentative steps the principle of the whole became visible all at once and we could add by pure deduction all the still missing elements.

Einstein's view of scientific research did not focus on physical discoveries applying in practice. Electromagnetic waves or laser beams and the like hardly interested him. He never looked seriously for practical physical knowledge, and barely for the profit which might result from it[66]. His goal was not the broad assembly of details but the depth of scientific knowledge. The scientist uncovering the great, unified and fundamental principles of nature from which all other insights then would follow was his ideal.

Einstein was not alone with that view in his time. Max Planck held similar notions, and we may safely assume that they arose from an understanding of science that was widespread at that period, at least in theoretical physics. Otherwise Einstein would hardly have taken so much for granted.

---

65 Fölsing, *Albert Einstein - Eine Biographie*, p. 627–643.
66 Fölsing, *Albert Einstein - Eine Biographie*, p. 446 ff., 450 ff.

III

For fully understanding Einstein's view we must consider it in its historical context. In early 20th Century many shared his notion that physics must be an 'exact science', and must therefore start from a secure basis and proceed from there *more geometrico* to new scientific insights. That notion expressed a belief inspired by the age-old fascination of Euclidean geometry. Einstein himself repeatedly invoked it, and many other physicists and mathematicians quite naturally presupposed it as being so obvious that there seemed to be no point in discussing it. Hardly anyone questioned it in his time[67]. I remind of his praise of Euclidean geometry:

> 'The man who was not enthralled in youth by this work was not born to be a scientific theorist."[68]

(1) That notion was deeply rooted in history and its origin lay in a misunderstanding. Since the 13th Century almost every philosopher standing in the Aristotelian tradition of scholasticism had assumed that scientific knowledge must stand on a reliably secured base. One believed that the unshakable edifice of science must have a solid foundation. 'Scientific' knowledge therefore required selecting the ground with diligence, laying brick by brick methodically and observing the rules of scientific craftsmanship with utmost care. The necessity of proceeding like that seemed so obvious that hardly anyone questioned it. One assumed that in order to obtain 'scientific' knowledge, every single step had to observe valid rules and must hence be 'justified'. As a result, for centuries to come, discussions of epistemology hardly concerned the process of obtaining knowledge any more, but rather the question of what foundation was to carry the edifice of science. To some it were the laws of logic, often also called

---

67   One of the rare exceptions was Henri Poincaré in *La science et l'hypothèse* (1902).

68   Einstein, *On the Method of Theoretical Physics*, Philosophy of Science, vol. (1934), p. 163 (quoted in *Chapter 1, III, 2*).

'laws of thought' and believed to establish knowledge *a priori*. To others it was experience or intuition. The impression of Aristotelianism in combination with Euclidean geometry was so overwhelming that most philosophers agreed at least on one point: only absolutely reliable and certain knowledge could serve as the foundation of 'scientific' knowledge.

René Descartes (1596–1650) in particular had shaped European philosophy in that sense. He himself had transferred the methods of algebra to Euclidean geometry and had developed a forerunner of modern analytic geometry excellently suited for describing physical processes. That remarkable success probably confirmed him in his belief that the axiomatic approach must apply universally. Seen from that angle his postulate seems natural that the starting point of our endeavours to obtain true knowledge must be *clara et distincta perceptio,* because only clear and distinguishable perception could be purified of all errors[69]. It also shows his axiomatic approach to the theory of knowledge. Like axiomatic geometry, the whole of empirical science was to rest on a foundation that was absolutely secure and indisputable.

Descartes' influence on European philosophy was immense. Even philosophers holding different views in almost every other respect yet followed him in his axiomatic approach. Thomas Hobbes (1588–1679), who knew him in person, was deeply impressed with that method. An approach step by step as demonstrated in geometry and mathematics seemed to him the very model of every science, and definitions seemed to be its foundations. Spinoza (1632–1677) even wanted to provide ethics with a foundation of axioms[70]. Leibniz (1646–1716) considered the rules of logic to be knowledge valid *a priori* from which all other knowledge of nature must emanate, explaining error thus by human deficiency alone[71]. Even Kant (1724–1804), the only philosopher to pursue a fundamentally different approach since about 1770, nevertheless believed that doing away

---

69   Descartes, *Meditationes*, III, IV.

70   Spinoza, *Ethica Ordine Geometrico Demonstrata* (posthumously 1677).

71   Leibniz, *Nouveaux essais sur l'entendement humain* (1704), vol. IV, ch. II, *Des degrés de nostre connoissance*; he distinguished there between ,Vernunftwahrheiten' (truths of reason) and ,Tatsachenwahrheiten' (truths of facts); also ch. XX, *De l'erreur*.

with all certain knowledge would be impossible, and thus retained in his *a priori* a last remnant of the axiomatic approach to which I shall return[72].

The great Anglo-Saxon empiricists Bacon (1561–1626), Locke (1632–1704) and Hume (1711–1776) took a different approach by assuming that not knowledge valid *a priori* but experience alone must be the foundation of knowledge. They, too, however, presupposed thereby that the knowledge on which science was to build must be reliably established (by experience), and they thus were caught in the logical turbulences of inductivism which we saw in *Chapter 2, I.* The difference of their starting point did not alter the principles of the method adopted. Even Karl Popper (1902–1994), himself so strongly impressed with Kant's theory, sometimes had difficulties in getting away from Cartesian thinking and noticed that himself[73].

The impact of the Cartesian approach on theoretical physics still shows in a paper on the *Meaning and Limits of Exact Science* which Max Planck read at Berlin in 1941. He raised the question there of how we might find a starting point for 'building up exact science' that would 'stand up to any criticism whatsoever', and answering it, went on to say[74]:

> 'In other words, we must focus our attention not on what we want to know but, first of all, on what we know for certain. – What, then, is the most certain of all we know and can impart to each other, subject not to the slightest doubt whatsoever? There is only one answer to that: is what we ourselves experience with our own bodies.' (My translation).

---

72 *Chapter 4, II.* – According to Popper's interpretation Kant was so firmly convinced of the truth of Newton's theory that he believed doubting it to be impossible. See Karl Popper, *The Nature of Philosophical Problems and their Roots in Science,* in Karl Popper, *Conjectures and Refutations,* p. 66, sec. x; also in *On the Status of Science and Metaphysics, Conjectures and Refutations* p. 184, 190 ff.

73 Karl Popper, *Beyond the Search for Invariants,* in *The World of Parmenides,* p. S. 146 ff. In the section on *Knowledge without Foundations* (p.152f.) he reported having been convinced only recently by Imre Lakatos that even mathematics has no 'foundations'. See also *Appendix 1, II.*

74 Max Planck, *Sinn und Grenzen der exakten Wissenschaft* in *Vorträge und Erinnerungen,* 5th ed. (1949), p. 363 ff.

That was – in the middle of the 20ᵗʰ Century and phrased by one of its most famous physicists – nothing else than the *clara et distincta perceptio* of Cartesian doctrine, a reference to the one clear, intuitive, distinguishable and unquestionable axiom on which all true knowledge was to build. The self-assurance showing in that statement is almost its most surprizing feature. Going by Planck's words it might have seemed that epistemology had stood still in the centuries since Descartes.

All this indicates how strong the influence of Cartesianism still was at the beginning of the 20ᵗʰ Century. Most scientists seem to have drawn the borderline between logic and mathematics on the one side and empirical science on the other side differently than we would draw it today. Distinguishing between a proposition and its object is a matter of course in modern science, not only when the object is empirical but also when it is abstract as for instance when we speak about language itself. But we owe that distinction mainly to Whitehead and Russell's *Principia Mathematica* (1910) and to Alfred Tarski's introducing the terms 'object language' and 'metalanguage' for discerning those different levels of language[75]. At the beginning of the 20ᵗʰ Century few saw that as clearly as they did.

When Einstein stated his special theory of relativity in 1905 and read the papers quoted in *Chapter 2, III*, the usual distinctions in German philosophy, traditional since Leibniz and Kant, were between 'analytic propositions' relating to the subject of a sentence and 'synthetic propositions' added to it in the predicate, and between knowledge *a priori* and *a posteriori*. Those distinctions depended more on the structure of a sentence than on the content of the statement. They were not intuitively accessible and hardly qualified for rousing anything like a general understanding of the difference between axiomatics and empirics. Besides, the distinction between knowledge *a priori* and *a posteriori* presupposed in both alternatives that certain knowledge is possible in principle.

Trusting in the axiomatic method thus was normal in Einstein's time. Scientists considered it right as a matter of course, particularly in physics

---

75 See for instance Tarski, *The Concept of Truth in Formalized Languages* in *Logic, Semantics, Metamathematics*, p. 152, 167.

which was believed to be the basis of empirical science. They thought that even if other disciplines, the humanities in particular, permanently lost themselves in the fog of speculation, at least in physics mere 'opinion' must not prevail but only statements belonging to 'exact science' be admitted. In order not to compromise that noble aspiration to exactitude, and probably also under the impression of Newton's belief in inductivism and his famous words *'hypotheses non fingo'*[76], one believed that physics must renounce all temptation to speculation and virtuously confine itself to what we really know for certain. Starting from secure foundations seemed essential for achieving that. One of the most famous physicists of the 19th Century, James Clerk Maxwell (1831–1879) who seems to have been the hero both of Planck and Einstein, was clearly recognizable as an adherent of the axiomatic approach by his style and the structure of his investigations, and by his belief in the possibility of proof in physics. There will be more of him in *Chapter 9*.

(2) In fact however, the axiomatic understanding of science had got in trouble long before Einstein's time. The invention of non-Euclidean geometries in the first half of the 19th Century had raised new questions. Sensitive natures may have felt that, even if they were not capable of articulating their qualms. Those questions must have caused a deep crisis in the self-confidence of many physicists.

Euclid had developed his system of geometry in the 3rd Century b. C. In the more than two millennia since his time, there had always been only one kind of geometry, and through the development of analytic geometry even that had merged with algebra and arithmetic into one comprehensive mathematical science. In every branch of it propositions could be obtained that were open to cogent proof. Under such circumstances it seemed plausible that the axioms of a science so convincing must be scientific *truths*, reliably secured therefore, and apt to serve as foundations for

---

76 See the quotation in *Chapter 1, I*. For Newton's belief in inductivism see also Rule III of his 'System of the World in Cajori, *Sir Isaac Newton's Mathematical Principles of Natural Philosophy and his System of the World*, appendix to vol.II, p. 398.

all other findings of that science. Geometry was considered the model of any 'exact science', and scientific physics had emerged from it[77]. For centuries to come, Galileo's statement of the laws of falling bodies, Kepler's discovery of the laws of planetary motion and Newton's even more general laws of motion had encouraged scientists in their belief that physics was an 'exact science' in the same sense as geometry. Exactitude was just what seemed to make out its scientific quality. Every academic physicist was also a mathematician, often primarily so. He would teach mathematics and geometry, and would use examples from classical mechanics and optics for demonstrating their practical worth.

Seeing the main import of physics in mathematical calculations seemed natural too. The most tangible progress brought about by Kepler's and Newton's theories showed in the almost incredible precision of their predictions, and demonstrating that presicion was most impressive from the point of view of didactics. They made solar eclipses predictable and lose their terror. In the further course of events they even led to the discovery of planets hitherto unknown, Neptune and Pluto[78]. It cannot come as a surprise therefore, if such achievements appeared to contemporaries almost as revelations. Physics thus mainly served as a means for demonstrating the usefulness of mathematics, particularly when taught to pupils of the upper social classes. It was considered an auxiliary science; primacy clearly belonged to mathematics. The self-esteem of any scientist believing in the axiomatic method but finding himself unable to meet its requirements of exactitude would have been in serious jeopardy. Under the rule of Newtonian theory the world of physics ostensibly was a well-ordered, closed world. It was a world in which the rules of mathematics, geometry and logic were believed to be supreme and eternal laws.

(3) The tranquillity of this idyllic world of mathematically minded physicists was shaken to its foundations in the first half of the 19th Century

---

77 Very detailed Popper, *On the Status of Science and Metaphysics*, in *Conjectures and Refutations* p. 184, 190 ff.

78 Neptune was discovered in 1846, Pluto in 1930.

when Carl Friedrich Gauss, Nikolai Lobachevsky and János Bolyai, largely independently from one another, developed spherical geometry and demonstrated the possibility of other non-Euclidean geometries. For the first time in history those great mathematicians based their 'alternative' systems on axioms quite different from those of Euclidean geometry. They particularly called into question the axiom of parallels, considered problematic already in Euclidean geometry due to its not being accessible to direct observation, and replaced it by other axioms. Starting from their new axioms they arrived at results quite different from those of Euclidean geometry. In their geometric systems the shortest distance between two points was no straight line, and not only one but several lines could be drawn through a point outside a straight line without cutting that line, and the angular sum of a triangle could be more or less than 180°. Based on those axioms they successfully demonstrated mathematical proofs as conclusive as those of Euclidean geometry were, and created axiomatic systems that were equally self-consistent. Yet their non-Euclidean geometries differed from Euclidean geometry not only in their axioms but also in their results.

That brought a fundamental change in the outlook on the basic questions of the theory of science. At least that is what should have happened. If several axiomatic systems could coexist independently, each of them self-consistent but all leading to different results, then the axiomatic understanding of science itself was called in question.

The axiomatic understanding of science had started from the notion that the axioms of science require no further justification because their truth is 'directly apparent', and that logical inference will transfer it automatically and without exception to any conclusion correctly drawn from them. The basic assumption, on which one tacitly relied without questioning it, was that the axioms must be *true* and that their truth must be *certain*. Science was a system of true statements in axiomatic understanding, and the truth of the axioms underlying that system was to guarantee the truth of all conclusions inferred from them as well as their self-consistency and mutual compatibility. Seen from that angle, error in science could only result from misapplying the rules of logic and mathematics. And since the truth of the axioms was assumed to be certain and 'directly apparent', error could be explained only as a symptom of personal deficiency or of dishonesty.

The notion that axioms are true in the sense of empirical truth became untenable however, when the non-Euclidean geometries developed by Gauss, Lobachevsky and Bolyai had shown that though based on different axioms, several axiomatic systems could nevertheless co-exist with equal rights and be equally coherent and self-consistent in spite of the different results following from them. The question of the truth of the axioms inevitably came up again after that because it could no longer be sufficient that the truth of axioms, such as that the straight line is the shortest line between two points, is 'directly apparent'. It is just as directly apparent, for instance, that in spherical geometry a straight line cannot be the shortest line between two points because spherical geometry has no straight lines. Hence it became necessary to make a choice between different axioms.

That meant, however, that the different geometries and their axioms stood in competition as it were. Science could no longer avoid the question of their justification. But how might they be justified? Even without justifying them it would nevertheless be necessary to make a choice between different axioms. What standards were to apply to that choice if the axiomatic method itself provided no criteria? That was one of the fundamental questions in the philosophy of science. Few saw it clearly but it was in the air nevertheless[79].

(4) Due to that question arising in the first half of the 19th Century anyone believing in the epistemological programme of a fully axiomatized empirical science should have given up that belief directly and completely. After the invention of non-Euclidean geometries science could no longer avoid the question of the truth of their axioms. But it sometimes takes long for digesting events of that kind.

---

79 One of the few seeing it clearly was Poincaré in *La science et l'hypothèse* (1902), p. 51–57.

The conflict between an axiomatic understanding of science admitting of only one system and the existence of various different axiomatic systems, equivalent but competing with one another, is irresolvable. Einstein never solved it; my impression is, rather, that he never noticed it[80]. In his writings I have found no other indication, and we have no reason to suppose that all scientists accepted at first sight an insight we now find convincing. Human beings are no computers. The road to gaining acceptance for scientific theories often is tedious, particularly when it comes to understanding and applying the more remote implications of a new idea.

We may realise today that the mere possibility of non-Euclidean geometries conflicts with the research programme of an axiomatized science but that does not mean that adherents of that programme would at once and willingly have given up their axiomatic understanding. The consequences resulting from the invention of non-Euclidean geometries did not reveal themselves automatically. Many scientists must have found it difficult to let go of old ideas.

In the early years of the 20th Century at any rate, when Einstein published his special theory of relativity, faith in the axiomatic method was still unbroken. It had even recently gained new support when David Hilbert succeeded in proving the self-consistency and mutual independency of the Euclidean axioms (1899) and thereupon undertook also the axiomatization of algebra[81]. At the turn of the century

---

80 Einstein believed that representing curved space required a non-cartesian system of coordinates (*Grundzüge der Relativitätstheorie* (The Meaning of Relativity) [1956], 6th ed., p. 63 f.). There is some logic in that, for if all space were curved, then plumblines and rulers too would be curved, and we therefore would be unable to draw straight lines. Nevertheless, the argument shows that Einstein confused mathematics and empirics. As every mathematician knows, describing curved lines in systems of straight coordinates raises no problem at all. Einstein must have been attributing to the respective non-Euclidean geometries empirical *truth* contents. I could find no direct indications for that in his writings, but the interpretation would agree with his general line of confusing mathematics and physics, of which we will see more in *Chapter 9*. See also my *Popper versus Einstein*, p. 79 f.

81 Hilbert, *Grundlagen der Geometrie* (1899 - Foundations of Geometry).

'a climate of opinion was thus generated in which it was tacitly assumed that each sector of mathematical thought can be supplied with a set of axioms sufficient for developing systematically the endless totality of true propositions about the given area of inquiry.'[82]

That was the intellectual climate which inspired Alfred North Whitehead and Bertrand Russell in their large-scale attempt at reducing the whole of mathematics to axioms of logic. In their *Principia Mathematica* (1910) their aim was to establish a comprehensive logicist system of non-empirical science. They carefully distinguished between a proposition and its object, and thus created the conditions from which Tarski later was to develop his theory of different levels of language, object language and metalanguage.

Like their contemporaries, Whitehead and Russell assumed mathematics and logic to be a unified system of internally consistent statements, and they believed an important task of science to consist in finding and spelling out a complete set of the axioms underlying that system. The idea that logic and mathematics might be no systems at all, but open in principle, still was alien to them. The general belief in logical or mathematical 'systems' was to crumble only after 1931 when Kurt Gödel had published his famous paper '*On Formally Undecidable Propositions of Principia Mathematica and Related Systems*'[83]. He proved there that even a simple system like that of the ordinary mathematics of integers cannot be provided with a set of axioms which is 'complete' in the sense that the totality of true propositions can be derived from them[84]. Moreover, he also proved that it is impossible to establish the internal logical consistency of complex deductive systems such as that of elementary arithmetic without introducing

---

82   Ernest Nagel/James R.Newman, *Gödel's Proof* (1959), p. 6.

83   Kurt Gödel, *Über formal unentscheidbare Sätze der Principia Mathmatica und verwandter Systeme* (1931 - On formally undecidable Propositions Of Principia Mathematica And Related Systems), Monatshefte für Mathematik und Physik vol. 38 (1931), p. 173–198.

84   I have read Gödel's proof only in the representation by Nagel/Newman, which I am following here.

principles of reasoning so complex that their consistency would be open to the same doubts as that of the system under investigation[85].

(5) Einstein never recognizably drew any conclusions for his own work from Gödel's proof. That is one of the strange inconsistencies in his life. He must have heard of it because at Princeton he was close friends with Gödel. Yet nothing indicates that he ever saw a connection between his own endeavours and Gödel's famous achievement. On the contrary, he even seems to have succeeded in drawing Gödel over to the side of the theory of relativity[86].

Why did Einstein, attaching so much importance to mathematics, never consider the implications of Gödel's proof? It need not have impressed itself on him directly that it was relevant for his theory of relativity. He did not believe the formulae of relativity to be fundamental axioms of science, but saw in them only forerunners to further knowledge still to be obtained. That was why he continued searching for the unified field theory and his 'world formula'. Yet how could he hope ever to succeed in that search after Gödel had *proved* the impossibility of axiomatizing completely even a system as elementary as the ordinary arithmetic of integers? Even the special theory already transgressed the narrow limits to axiomatization thus revealed, to say nothing of the far more complicated general theory.

Why did Einstein continue his search for the basic axioms of nature? Should he not have given up his quest? That remains a mystery. Going by his own theory of science quoted in *Chapter 2, III, 1*, he needed an additional axiom for transposing into physics the rules of Euclidean geometry. Even that already meant that Hilbert's proof did not apply to the enlarged system. And how could the self-consistency of a theory as complex as the general theory of relativity be established mathematically after Gödel had

---

85   Ernest Nagel/James R.Newman, *Gödel's Proof* (1959), p. 6.

86   Gödel was already seriously ill at that time. For details see Yourgrau, *Gödel, Einstein und die Folgen* (2005 - A World Without Time - The Forgotten Legacy of Gödel and Einstein), p. 120 ff. Yourgrau does not mention that Einstein had drawn any inferences for his own work from Gödel's proof.

shown that the proof would then have been open to the same doubts as the theory itself?

Such questions come to mind but Einstein seems not to have considered them. There is no indication in his writings that he ever tried to think about the logical implications following from Gödel's proof for his own field of research. He must simply have overlooked the fact that they refuted not only his physical theories but the whole approach he had taken. And no theoretical physicist after him seems to have noticed that either.

CHAPTER 4:

AXIOMATICS VERSUS HEURISTICS (1): FROM KANT TO EINSTEIN

Einstein's approach was Popper's inspiration. I. Historical background; revolutionary developments of science in 19ᵗʰ Century; new problems raised by discovery of electricity; Maxwell vs. Faraday; unconditional trust in the truth of mathematics and ether theory both generally accepted in 19ᵗʰ Century. II. Einstein's promise to do away with ether and return to 'exact science'; the two premises of his special relativity; originality of his heuristic method almost unnoticed; its revolutionary aspect: premises are 'free inventions'; reversing the direction of cognition; permitting creativity in science; putting personal responsibility on the scientist. III. The origin of Einstein's approach in Kantian theory; my personal interpretation of Kant; his Copernican revolution; his anticipation of Popper's criterion of falsifiability; Einstein's axiomatic approach to physics inspired by Kant's reference to the 'formal unity of nature'; Kant's objectivism and realism; time aspect of his principle of autonomy; its control by experience; its relevance for ethics.

Einstein's thoughts about the theory of science were not confined to the axiomatic approach. His papers reveal also another principle at least equally important for understanding him. It is in his *heuristic method*, and I consider it his greatest achievement. It gave Karl Popper the inspiration for his own theory of scientific discovery. Heuristic method would prove extremely useful for theoretical physics if applied consistently, but this feature of Einstein's thoughts was so revolutionary that few seem to have understood it in his time.

I

The 19th Century in which Einstein grew up had been a brilliant era for science in many respects. In number as well as in importance the discoveries made then were without precedent in the history of science. They did not concern physics alone but covered literally all fields of life. The whole world looked on science and became accustomed to expecting from it new sensations almost day by day.

(*1*) There always are two sides to a medal however. The progress of science in the 19th Century was spectacular indeed but the period need not have been a happy time also for every scientist living in it. It makes a difference whether success is yours or you watch that of others. Scientists less gifted with creative imagination may have felt bowled over by developments rather than carried away by them. Witnessing others coming to reputation and wealth while they were fighting for subsistence might have roused quite different feelings. And such feelings, too, can make an impact if shared by many.

The intellectual climate of the period must have been an important factor in the development of science in the 19th Century. The French Revolution of 1789 still was in everyone's mind then. It had originated in economic and social strife primarily, but it was also the expression of a new mental attitude felt far beyond France's borders. Not envy stood at the forefront but justified discontent with those who by birth and rank should have taken on responsibility and instead rested on inherited wealth and basked in unmerited privileges. In France the destructive element of the revolution prevailed temporarily. Yet there too the new state of mind was no sign of weakness but of a fundamental change in the balance of power in society, and of feeling able to do things better. That explained the resonance which the ideas of the French Revolution found all over the world. The bourgeois had become aware of their own worth and self-confidently placed themselves on a level with the previously ruling classes of the nobility and the clergy, increasingly often even above them. The value of the individual was beginning to separate from the social class from which it came.

In the natural sciences that development contributed to the optimism mentioned above. Together with other factors it created an intellectual climate as we perhaps experience it again in our time in the development of electronic data processing. The whole world was full of explorers, inventors and visionaries, not always easy to distinguish from serious scientists. Rise and fall often came close together because no one knew what would turn out to be right in the end. Experimenting was important therefore, and everything was freely considered and discussed.

The world thus entered into an era of physical discoveries which jumbled up the hitherto so lucid and well-ordered system of classical physics. The discoveries about electricity in particular came so fast that even the categories needed for *describing* them still needed to be developed. In our time we know not only that electricity exists but also how to use it. In early 19th Century however, its parameters, such as electric tension, electrical charge, flowing current and resistance, all were unknown still, and neither standards existed for describing them nor instruments for measuring them[87]. In the face of those phenomena the theoretical tools of classical physics which had survived still unscathed the discoveries made by Copernicus, Kepler, Galileo and Newton, failed all along the line. Events simply passed over them.

That showed already in the way that era began. The first to pay special attention to the phenomena of electricity, Luigi Galvani (1737–1798) and Franz Mesmer (1734–1815), were neither mathematicians nor physicists. The were medical men, therefore belonging to a profession which had been despised as 'unclean' still at the beginning of the 17th Century. Their scientific understanding, if any, was almost directly opposed to that of

---

87 That is part of my family history. My ancestors Rudolf Kohlrausch (1809–1858) and his son Friedrich Kohlrausch (1840–1910) were deeply involved in developing methods and standards for measuring electricity in its various manifestations. See for instance Wilhelm Weber/Rudolf Kohlrausch: *Über die Elektrizitätsmenge, welche bei galvanischen Strömen durch den Querschnitt der Kette fließt* (1856), in *Fünf Abhandlungen über absolute elektrische Strom- und Widerstandsmessung*, ed. by Friedrich Kohlrausch (1904).

mathematicians. In view of such developments a scientist having mentally fled from the cultural and economic upheavals of the time to the safe harbour of classical physics must have felt serious apprehensions.

The old social system with its feudal order had not put mathematicians and physicists in a top rank of social hierarchy, yet had assigned to them a respectable place appearing also reliably secured. Though promising no wealth, it at least ensured their belonging to the spiritual republic of scholars and thus creatd a notable distinction from common people and a tolerable social standing. The technological revolution of the 19ᵗʰ Century swept away that order of classes too. In those times of general upheaval even scientists would no longer be measured by their rank or their learning but quite profanely by the results they achieved. Revolution took place not only in the political sphere. It permeated the social fabric of the old world in all its ramifications, stopping not even at mental issues. At the level of science it was a revolution against the system of classical physics itself. By the second half of the 19ᵗʰ Century at the latest the security of the old order had broken down there too.

Scientists finding their ideal in classical physics and hoping to preserve it, might well have despaired of such developments. Nameless upstarts lacking any academic education whatsoever let their speculations blithely run wild and even came to fame like that. Michael Faraday (1791–1867) possibly was the extreme case. We know today that modern science and technology owe to him almost everything. In the 19ᵗʰ Century however it could still seem presumptuous that a mere bookbinder's apprentice and son of a blacksmith should actually have read the books he was only to bind or to repair. His knowledge, too, which he had acquired by self-studies, initially was more about chemistry than about physics. He knew little about mathematics and even had the cheek frankly to admit that[88]. His

---

88 In 1822 Faraday wrote in a letter to Ampère: 'I am unfortunate in a want of mathematical knowledge and the power of entering with facility into abstract reasoning. I am obliged to feel my way by facts closely placed together, so that it often happens I am left behind in the progress of a branch of science (not merely from the want of attention) but from the incapability I lay under of following it, notwithstanding all my exertions. It is just now so, I am ashamed to say, with your refined researches on electromagnetism or

highly speculative hypotheses about invisible radiation and latent fields of forces showed physical intuition but they raised no claim even to the slightest deductive foundation. Yet his experiments repeatedly confirmed those speculations. His discovery of electrical induction (1831) was widely recognized although it not even pretended to any kind of deductive reasoning. In view of such visible signs of decay many an academic physicist clinging to the ideal of an exact science must have been tearing his hair. Some may even have decided to turn their backs on practical physics altogether, and to devote their time to pure theory alone in order to remain undisturbed henceforth. That may have been why James Clerk Maxwell (1831–1879) saw an important task in bringing Faraday's reckless speculations back to the secure foundation of mathematical deduction as quickly as possible [89]. The gulf between experimental physics and theoretical physics was beginning to grow into the 'schism in physics'.

(2) At the turn of the 19th Century one of the features of the intellectual climate of science was an exaggerated trust in mathematical deduction. The unprecedented power of persuasion of geometry had long fostered it, and it endowed many scientists with an unshakeable belief in the universality of the axiomatic method and in the flawless predictability of physical processes.

We saw in *Chapter 3, III, 3* that non-Euclidean geometries had been invented in the first half of the 19th Century. That should have raised the question whether mathematics and geometry really are expressions of empirical truth or whether they are non-empirical. In fact however, most scientists of the 19th Century seem to have interpreted that invention merely as supporting their own personal view that to every problem of physics there must be approaches, and ultimately even solutions, at the

---

electrodynamics. On reading your papers and letters, I have no difficulty in following the reasoning, but still at last I seem to want something more on which to steady the conclusions.' (Quoted from Peter Day, *The Philosopher's Tree, A Selection of Michael Faraday's Writings*, p. 96 f.).

89 For Maxwell's approach to the problems of electricity see *Chapter 9, I* and *VI, 1c*, below.

level of mathematics. Maxwell's *Treatise on Electricity and Magnetism* (1873) seems to have encouraged that belief, and Hilbert's proof of the self-consistency and mutual independency of the Euclidean axioms (1899) reinforced it. Whitehead and Russell had not yet failed in their attempt at reducing to the axioms of logic the whole of mathematics. Belief in the unlimited possibilities of the axiomatic method was still unbroken therefore.

Parallel to that, however, there lived in the general outlook of theoretical physics at the turn of the 19th Century also another basic assumption, incompatible with the one just mentioned, yet also widely accepted as a matter of course. That was the ancient *ether hypothesis*. Discovering physical effects that are normally hidden to human senses had become common experience of scientists by then. That space is not empty even where no matter can be observed was believed to be about as self-evident as the Big Bang theory is believed to be self-evident by many in our time. Henri Poincaré and Hendrik Antoon Lorentz among many others were staunch supporters of the ether hypothesis. They assumed the universe to be filled with an omnipresent invisible substance serving as the carrier of light.

In late 19th Century belief in the axiomatic method and belief in ether theory were both widely spread. They mostly were so nondescript that they hardly deserved being considered 'theories' at all, at least not scientific theories in any strict sense of the word. Rather, they were in the way of intellectual undercurrents, or speculations, occasionally turning up here or there but without being stated precisely. Thus they could live peacefully side by side for a long time without interfering with one another. Max Planck accepted the ether hypothesis but also relied on the axiomatic method and never recognizably took umbrage at the conflict between them. Most of his contemporaries did likewise. Einstein seems to have been the only scientist wo recognized a serious problem in the coexistence of the two worlds of thought based on those ideas.

However, even if most physicists accepted ether theory at that time that does not mean that they liked it. Some of the more mathematically minded among them must actually have hated it, especially if they held it responsible for the problems of mathematical astronomy raised by the unknown velocity of light in outer space.

II

Into that tense atmosphere came the publication of Einstein's light quantum hypothesis, of his special theory of relativity and of his famous formula $e = mc^2$ postulating the equivalence of mass and energy, to mention only three of the papers he published in 1905, his so-called *annus mirabilis*[90]. We will see in the following chapters that none of them carried conviction in itself. That is my chief reason for believing that the intellectual climate into which they came must have strongly contributed to their success.

In the second of those papers, headed *Zur Elektrodynamik bewegter Körper* and containing his special theory, Einstein claimed that accepting only two premises would open up a new approach to the theories of moving bodies and of electrodynamics. One of them was that the laws of nature are valid throughout the universe, and the other was that the vacuum speed of light is constant, independent of the motion of its source. He promised that those premises would permit to solve all problems in these fields without resorting to ether theory.

That promise gave new hope to scholars of the old school because it meant returning to an understanding of science in which physics would once again yield absolutely certain and unquestionable knowledge. Like Euclidean geometry it was to start from axioms and to proceed from there by purely mathematical inference to new insights that would then be as certain as the axioms themselves. Einstein thus undertook to give back to the community of academic physicists their lost self-confidence and to lead them into the Promised Land of *exact science*. With that he could rest assured of the benevolence of all those having committed themselves to the scientific ideal of classical physics.

---

90 Albert Einstein, *Über einen die Erzeugung und Umwandlung des Lichts betreffenden heuristischen Gesichtspunkt* (On a Heuristic Point of View Concerning the Production and Transformation of Light), Annalen der Physik vol. 17 (1905), p. 132 ff.; *Zur Elektrodynamik bewegter Körper* (On the Electrodynamics of Moving Bodies), Annalen der Physik vol. 17 [1905], p. 891 ff.; *Ist die Trägheit eines Körpers von seinem Energieinhalt abhängig?* (Does the Inertia of a Body Depend on its Energy Content?, Annalen der Physik, vol. 18 (1905), p. 639.

(*1*) It must have been due to those enticing prospects that another feature of Einstein's approach remained almost unnoticed at the time. That was the heuristic method which he employed in his papers.

I cannot say whether he had recognized in it a methodological principle from the outset, or whether others roused his interest in it when they asked him to speak about the method he employed. At any rate he later commented on his physical theories from the angle of epistemology as we saw in *Chapter 2, III*. His reflections set in at the question of the validity of the axioms and he answered it with a proposal as radical as it was revolutionary and different from anything previously proposed. I repeat that in my view it was his most important contribution to epistemology. It gave Karl Popper the decisive inspiration for his theory of scientific knowledge.

Before Einstein, theorists of physics had always been searching for a reliable basis of knowledge. Their aim had been to establish a secure foundation on which all further knowledge could build, and all had foundered on that rock. Unlike his predecessors however, Einstein not even attempted to justify in any way the axioms he proposed. Instead, he summarily declared them to be

> 'free inventions of the human mind which admit of no *a priori* justification either through the nature of the human mind or in any other way at all.'[91]

He even explicitly spoke of 'assumptions' which he then raised to the rank of 'premises' in his papers[92]. He gave no further justification. Only the conclusions he drew from those premises were to convince his readers. They were his sole argument.

---

91   Albert Einstein, *Zur Methodik der theoretischen Physik* (On the Method of Theoretical Physics), p. 185 ff., 189. For the context see *Chapter 2, III, 2*, above.

92   Albert Einstein, *Zur Elektrodynamik bewegter Körper* (On the Electrodynamics of Moving Bodies), Annalen der Physik vol. 17 [1905], p. 891.

Instead of constructively establishing the basic truths of science by deriving them from other propositions and placing them on a solid foundation thereby, as others had tried in the past, Einstein introduced his ideas as completely unfounded allegations at first, and proposed to examine their truth only subsequently by comparing them to experience. They were random only seemingly however, because they had to be possible solutions to existing problems. That was their 'justification', coming not before but after introducing them.

That was Einstein's often mentioned heuristic approach which he took not only in his paper on the special theory but in others too[93]. It particularly shows in his light quantum hypothesis, which we will discuss in *Chapter 8*. He would begin by stating seemingly arbitrary hypotheses such as for instance those of the quantum nature of light or of the constancy of the speed of light, without deducing them from other findings or justifying them in any other way. Instead, he would only claim at first that they were possible solutions to open problems. Deciding on whether or not they were correct solutions was to depend on their subsequent examination at the hands of experience.

(2) Seen from the angle of epistemology the originality of Einstein's approach can hardly be overrated. It turned the theory of knowledge upside down as it were, because it proposed reversing the direction of what had until then been regarded as the process of human cognition.

Before Kant (1724–1804) all philosophers approaching the problems of epistemology had started from the notion that the way of obtaining knowledge of nature must lead *from nature to man*. Nature would speak to man, and man would be her attentive but humble and passive audience, taking in her revelations and partaking thus of their truth. That conformed to an ancient tradition probably going back to Plato and Aristotle, and no philosopher but Kant had questioned it in millennia. Even the

---

93 For the revolutionary aspect of Einstein's approach at that time see Fölsing's *Einstein - Eine Biographie*, p. 158 f. - Elie Zahar claimed to be discussing that aspect in *Einstein's Revolution - A Study in Heuristics* (1989) but gave no information on its relevance.

great Anglo-Saxon empiricists believing knowledge to begin with observation, thought that observation must be prior to knowledge. The direction of the process of cognition was the same in their view, going from nature to man, not from man to nature. The more observations they had made the better would be their knowledge. Understanding nature is a *receptive* process in that view. Nature embodies truth, and man partakes of it if his good fortune has placed him among the elect to whom it was given to see it.

In a theory based on this approach the image of the scientist or of the philosopher will be one of detached objectivity. Due to his profession he is vastly superior to common people. And for preserving that superiority he must keep sufficient distance to things and to people, in short distance to everything he observes and investigates. He must be of pure mind, and must beware of jeopardizing his scientific objectivity and of clouding his judgment by contingencies or even by moods. The image of the scientist or philosopher underlying this view is that of a man distinguished from others by his aloof position and by his objectivity. His knowledge differs from that of ordinary people not only in quantity but also in quality. He not only knows more, he knows better. His knowledge is more precise, deeper and more certain because it is 'scientific' knowledge. That notion existed in nearly all centuries, and it still is widely spread in all branches of science[94].

Einstein's heuristic approach amounts to the opposite. It does not call in the scientist for detached objectivity but will allow him to be enthusiastic and passionate, if possible even ingenuous, leaving him thus room for genuine creativity. He may use 'intuition based on empathy in experience.' At the same time however, it confers on him a responsibility. And it is a personal responsibility indeed, resting not in nature but in himself, because he *decides* on the approach he takes. His passion for truth thus is

---

94 The tradition may have originated in Plato's theory of ideas. Francis Bacon's *Novum Organon Scientiarum*, assuming two ways to truth (aph. 19) and distinguishing between ideas and idols (aph. 23), is characteristic of the approach. Ideas come from 'the divine'; they show 'the true signatures and marks set upon the works of creation as they are found in nature.' Idols, on the other hand, 'so beset men's minds that truth can hardly find entrance'; man must therefore 'fortify' himself 'against their assaults (aph. 38).

the counterpart of his responsibility for truth. In Einstein's theory like in Kant's and Popper's, obtaining knowledge is no passive process leading from nature to man, but an *actively creative* process leading in exactly the opposite direction from man to nature. Scientific understanding of truth no longer is an outside event to which man is exposed in that notion. It lies in his own autonomous power. He *creates* the theories of science; they are his own achievement. And only after having created them can he examine them with respect to their being suited for solving a problem or to their conforming to experience. That makes out his responsibility.

Einstein himself thus first invented the hypothesis of the quantum nature of light and then investigated what problems it would solve. The hypothesis of the constancy of the speed of light was not his invention because it must have existed before he wrote his paper on special relativity. Nevertheless he introduced it there without attempting to justify it in any way, as a 'premise' on which he would base his theory. And he showed only afterwards how it was to be reconciled with experience. That is why others should thereupon have put it to the test.

The notion of science as an edifice jointly erected by all scientists carefully setting stone by stone was incompatible with that approach. In Einstein's understanding the scientist no longer was a craftsman carefully laying the bricks of knowledge according to a foreign plan. He could even believe himself to be the architect designing the building.

III

Einstein's heuristic method was a direct sequel to Kant's approach to epistemology. That relationship deserves closest attention here because it is the most important parting of the ways in the theory of knowledge, a point where opinions easily separate, particularly in physics. I will often have to refer to it in the further course of this essay.

Almost every fallacy in the theory of science was and still is connected somehow with implications of Kant's principle of autonomy which had either not been understood or not been properly applied. I am telling of personal experience by saying that. Though long convinced in principle by Kant's approach, realizing its implications nevertheless took me more than ten years in one case, and in another case the same happened only while working on this book[95]. Karl Popper had similar problems as we will see in the next chapter and in *Appendix 1*. Even Kant himself did not consistently apply his own theory to the natural sciences. Being misunderstood so often was his own fault to a large extent.

I will try to illustrate Kant's thoughts with quotations from his writings but must warn readers that his style does not make this an easy task. My reason for quoting him *verbatim* is not providing a reliable interpretation. The quotations express, rather, what I believe to be important in his ideas. My selection does not aim at objectivity therefore, but at representing Kant's theory as I understand it, and in a shape which I believe to be the strongest possible. It might thus show also how Einstein understood him.

(*1*) Kant's *Critique of Pure Reason* (1781) was one of the great events in the history of ideas. In that treatise he first explained what he later proudly called the 'Copernican revolution' in his theory of knowledge. By alluding to that most revolutionary cosmological discovery ever in the history of the natural sciences he wanted to draw attention to the fact that his own theory was equally revolutionary. As Copernicus had turned cosmology upside down so was he turning epistemology upside down. The revolutionary aspect was important to him. It deserves close attention therefore.

Before Copernicus most people had believed in Ptolemaic theory. The earth was at the centre of the cosmos and the sun and the planets orbited her. That view had prevailed unchallenged since antiquity. It was common

---

95 The first case concerned legal theory. During my time at the London school of Economics in 1969/70 my intention had been to demonstrate its connection with Popper's theory but I had to give up that plan then because I found myself going in circles. I found the approach which I believe to be correct only about ten years later and published it in *Recht und Rationalität* (1984). It is now also (in English) in my paper on *The Problem of Objectivity in Law and Ethics* in *Popper's Open Society after 50 Years* (1999). – The other case mentioned in the text concerns the laws of conservation. I will explain it in *Chapter 11, VI (3)*.

to almost all religious creeds and it agreed with observations people can make day by day. In clear weather anyone can see with their own eyes that the earth is standing still and that the sun rises in the East and sets in the West. Yet Copernicus stood this ostensibly certain knowledge on its head by placing the sun in the centre of the universe and letting the planets orbit him. He even placed the earth among those planets, moving her out of the centre thus and banishing her to periphery. Galileo gave a forceful account of the revolutionary impression that made in his time. In one of the dialogues of his *Sidereus Nuncius* he let his proponent of Copernican theory say:

> 'I shall never be able to express strongly enough my admiration for the greatness of mind of these men who conceived this (heliocentric) hypothesis and held it to be true. In violent opposition to the evidence of their own senses and by sheer force of intellect, they preferred what reason told them to that which the sense experience plainly showed them … . I repeat there is no limit to my astonishment when I reflect how Aristarchus and Copernicus were able to let reason conquer sense, and in defiance of sense make reason the mistress of their belief.'[96]

Kant was comparing his own revolution of epistemology to that development. He, too, was about to question what was widely seen as a matter of course and believed to be absolutely certain. As Copernicus replaced the geocentric cosmological view of antiquity by his heliocentric system, so he replaced both the theocentric epistemological view of antiquity and the nature-centred view of Anglo-Saxon empiricism by his *anthropocentric* theory of the autonomy of the human mind. Man was no longer receiving orders from nature. He himself was to be the centre of cognition, confronting nature with his own self-made theories. That was the revolution of epistemology which Kant wanted to bring about.

---

96 Galileo Galilei, *Siedereus Nuncius*, in *Galileo Galilei, Schriften, Brief Dokumente*, vol. I, p. 288. The translation is from Popper's *The Myth of the Framework*, p. 85.

(2) But although their approaches were running on parallel lines, the task Kant had set himself was more difficult even than that of Copernicus. All arguments were on his side but similar success was not to be his. In the field of abstract ideas clear understanding may sometimes be more remote from our minds than the planets are from the sun. That was why Kant met with even greater difficulties in expressing what he wanted to say, and in making others understand it. Even his own texts are not free of contradictions. With good reason Popper was to say of him 'But in vain did he protest. His difficult style sealed his fate.' [97] Let us nevertheless take a look at Kant's own words:

> 'Exaggerated and incongruous though it may seem to say that *mind is itself the source of the laws of nature*, and therefore of *the formal unity of nature*, that claim nevertheless is correct, and appropriate to experience. It is true that *empirical laws as such can by no means derive their origin from pure understanding*, as little as the immense variety of phenomena can adequately be understood from the pure form of intuition through our senses. *But all empirical laws are only specific provisions of the pure laws of reason*, under which, and by the standards of which, they only become possible; and the phenomena assume their lawful aspect; just as all phenomena, notwithstanding differences of their empirical aspects, must at any time conform with the conditions of the pure form of the senses.'[98] (My translation and italics).

My interpretation of these words is this. By claiming mind to be 'itself the source of the laws of nature' Kant wanted to say that other than hitherto believed he did not consider the process of finding truth as coming from an external influence on man, but as an autonomous and creative performance coming from man himself. His hinting that 'empirical laws as such can by no means derive their origin from pure understanding' might indicate a borderline between empirical and non-empirical statements. In

97 Karl Popper, Kant's *Critique and Cosmology*, in *Conjectures and Refutations* (1963), p. 179.
98 Immanuel Kant, *Kritik der reinen Vernunft*, p. 181.

another context Kant even spoke of the 'touchstone of believing to be true', anticipating in outline thus what Popper later was to establish as his criterion of falsifiability[99].

On the other hand Kant's reference to the 'formal unity of nature' probably indicated that he too, at least temporarily or in some wording that may have particularly impressed Einstein, assumed the possibility of a self-consistent axiomatic *system* of science. His belief that 'all empirical laws are only specific provisions of the pure laws of reason' can in substance be taken as a generalization of Einstein's notion mentioned in *Chapter 2, II* that the non-empirical axiomatic science of geometry could be converted into the empirical axiomatic science of physics by adding to it another axiom .

Some of the fallacies we saw in Einstein's thoughts were thus to be found with Kant already. Popper saw that whereas Einstein never commented on it[100].

(3) We must stay with Kant's principle of autonomy for another while in order to show its universal relevance. In the preface to the second edition of his *Critique of Pure Reason* Kant put even stronger emphasis on the idea of the autonomy of the human mind, expressing it also more clearly. He carried it to the extreme there by postulating that not knowledge must conform to nature but on the contrary, nature must conform to knowledge. That statement, which might have given Oscar Wilde the inspiration for his essay *The Decay of Lying*, may seem absurd at a first glance but it deserves being taken seriously. Kant's words were these:

> 'One previously thought that all our knowledge must be taken from the objects; but all attempts at making out about them something *a priori* through concepts, which would expand our knowledge, went to naught under this assumption. It is worth trying

---

99 Immanuel Kant, *Kritik der reinen Vernunft*, p. 688; Karl Popper, *The Logic of Scientific Discovery*, chapter IV (p. 78–92).

100 For Popper's view see for instance his *Kant's Critique and Cosmology*, in *Conjectures and Refutations* (1963), p. 180 ff.

therefore, whether we might not make better progress in the tasks of metaphysics by assuming *that the objects must conform to our knowledge* ... It is here as with the first thoughts of Copernicus who, after it had not gone well with the explanation of celestial motions when he assumed the entire host of stars to be revolving around the spectator, tried whether he might not be more successful by letting the spectator revolve and leaving the stars at rest.' [101] (My translation, second italics mine).

For getting the meaning of that, we must bear in mind that reality took on a kind of controlling function in Kant's world of thought. It was the touchstone of truth for those laws of nature which human mind had invented.

Kant neither was idealist nor subjectivist in the sense that he wanted to concede the epithet of being 'real' only to the inner notions of man. Idealism in the sense of taking objective material reality to be unprovable, or imaginary, or even just doubtful, he rejected explicitly and very clearly[102]. When suggesting that the objects must 'conform to our knowledge' he did not mean any physical influence of knowledge on reality therefore, but simply *priority* in the strict sense of the word, that is *in order of time*. All he wanted to say was that our own brain *creates* the image we have of nature, and that we form thoughts and ideas first, and only then, subsequently, can examine their relation to reality. We begin by forming expectations and find out only afterwards whether or not they are true.

> 'For the fact that the concept *precedes perception* signifies only its possibility; perception however, giving substance to the concept, is the only character of reality.' [103] (My translation and italics).

101 Immanuel Kant, Preface to 2nd ed. of *Kritik der reinen Vernunft*, p. 25.

102 Immanuel Kant, *Kritik der reinen Vernunft*, Transzendentale Analytik', Widerlegung des Idealismus, p. 254.

103 Immanuel Kant, *Kritik der reinen Vernunft*, Transzendentale Analytik', Postulate des empirischen Denkens, p. 253.

In that passage the temporal aspect shows clearly in the expression that concept 'precedes' perception. It also indicates that Kant saw between perception and external reality a strong correlation established by the law of cause and effect, which he considered a synthetic judgment that must be valid *a priori* [104]. That correlation served also as a controlling instance in his view. For explaining it he argued:

> 'If, however, we tried to shape from the material, which *perception* provides, new concepts of substances, of interactions, without borrowing from experience itself the example of their connection: we would *get into mere fantasies* the possibility of which would have no indication whatsoever speaking for them, because they would neither accept experience as their teacher, nor be borrowed from it.' [105] (My translation and italics).

Kant thereby expressed that concepts will indeed precede 'perception' but must be controlled by it because otherwise we would 'get into mere fantasies'. Controlling them must come from perception, which is but another word for experience.

(4) In his *Critique of Pure Reason* Kant applied this principle of the intellectual autonomy of man only to the theory of knowledge of nature. Later however he consistently transferred it also to ethics. His moral philosophy is beyond the scope of this essay but I must quote two short passages here in order to show the universal significance of the idea. In his *Foundation of the Metaphysics of Morals* Kant said:

---

104   For the temporal aspect, see Kant, *Kritik der reinen Vernunft, Transzendentale Analytik, Grundsatz der Zeitfolge nach dem Gesetze der Kausalität*, p. 227. The temporal interpretation is important for the connection with Popper's theory because Popper also often emphasized that theory must be prior to observation. See, for instance, Karl Popper, *Eine Welt der Propensitäten* (A World of Propensities) p. 59, 84, 89.

105   Immanuel Kant, *Kritik der reinen Vernunft, Transzendentale Analytik, Postulate des empirischen Denkens*, p. 251.

'It would be impossible also to give worse advice to morality than by borrowing it from examples. For, every example of it that is given me must itself first be considered by standards of morality as to whether it is worthy of serving as an original example, that is as a model; but by no means can it first confer the supreme notion of it. Even the saint of the Gospel first must be compared with our ideal of moral perfection before being recognized as such.'[106] (My translation).

Moral laws and ideals, too, were autonomous creations of the human mind in Kant's theory. The aspect appeared so important to him that in his treatise on *Religion within the Limits of Pure Reason* he finally almost exaggerated it by saying:

'It may seem precarious but is, in fact, in no way objectionable to say that every man makes himself a God, indeed, by moral standards (...), must make one himself in order to worship in him his creator.' (Popper's translation).[107]

That sentence, reminding of Lessing's *Ring Parable*[108], was no blasphemy. Kant was a religious man but he wanted to demonstrate that regardless of creed the principle of autonomy must apply universally and without any restriction. It simply follows from the fact that our senses are physical instruments and that our perceptions as well as our thoughts are the results of physical or chemical processes ending in our brain. Our senses will receive stimuli from outside, mostly in the shape of waves of various types and frequencies to which they react. And our brain will transform those stimuli into visualizations or into acoustical or other impressions. Only

---

106 Immanuel Kant, *Grundlegung zur Metaphysik der Sitten* (Groundwork of the Metaphysics of Morals), *Übergang zur Metaphysik der Sitten*, p. 36.

107 Immanuel Kant, *Die Religion innerhalb der Grenzen der bloßen Vernunft, Vom Afterdienst Gottes in einer statutarischen Religion* (Religion within the Limits of Reason Alone), p. 839; Popper quoted the passage in *Immanuel Kant, the Philosopher of Enlightenment, Conjectures and Refutations* (1963), p. 182.

108 Lessing had published his *Nathan the Wise* in 1783.

the results of those involuntary and unconscious transformations come to our conscious mind. Processing those stimuli in our brain will always be prior to our conscious impressions, and examining them by comparing them to empirical or moral standards will come only afterwards. Kant thus already indicated the close connection between science and responsibility which Karl Popper later adopted and to which I will return.

CHAPTER 5:

AXIOMATICS VERSUS HEURISTICS (2): FROM EINSTEIN TO POPPER

'Yet the superstition
In which we have grown up, not therefore loses.
When we detect it, all its influence on us.
Not all are free that can bemock their fetters.' [109]

GOTTHOLD EPHRAIM LESSING (1729–1781)

I. Inconsistency of Kant's a priori; applying the principle of autonomy to language and logic; Popper's inconsistency in applying methodological nominalism. II. Alfred Tarski: his method more important than his definition of 'truth'; his theory of logic; definitions of simple terms; all rules of logic reducible to ne venire contra factum proprium; the subjective element in Kant's principle of autonomy; its connection with Popper's critical rationalism; standard for measuring logic is expediency; Popper strongly influenced by Einstein; his ambiguity on the role of logic; his belief in 'axiomatic systems'; his connecting 'systems' with objectivity a remnant of Einstein's 'exact science' and Kant's a priori. III. Einstein's inconsistency: giving up searching for justifications but not searching for axioms; comparing Einstein and Popper; Popper's criterion of falsifiability as a demarcation of empirical science; Einstein's step beyond Kant by rejecting a priori; his inconsistency in applying it.

In spite of his belief in the autonomy of the human mind Kant did not consistently apply the principle he had discovered, and his inconsistency arose from his understanding of the natural sciences. Karl Popper expressed that most clearly in his Preface to the second edition of his *Logik der Forschung* (1966) where he wrote:

---

109   *Nathan the Wise*, IV, 2 (William Taylor's translation).

'Kant believed that there is a 'pure natural science', synthetic and at the same time *a priori valid*, and therefore *certain*. He believed that because he rightly saw that (1) Newton could not have based his physics on a collection of statements of observation. He also believed, probably inevitably in his time, (2) that Newton's physics was true. Together, those two theses yield Newton's physics being valid *a priori*, as claimed by Kant, for instance, in his *Metaphysische Anfangsgründe der Naturwissenschaft* (1785). We learned from Einstein, however, that Newton's physics might be wrong; and that means a complete change of the problem situation compared to the one Kant had been facing. (My translation, Popper's italics).[110]

Yet not even Popper recognized all the implications of Kant's principle of autonomy. Newton's theory might be wrong of course. But if it were so, then that would have nothing to do with Einstein's theory. It would follow from the fact that Newtonian theory is an empirical theory and therefore open to refutation by experiment. Einstein's special theory however is self-contradicting as we saw in the *Introduction*. And it is non-empirical as we will see in *Chapter 9*.

Popper saw that we invent the laws of nature and that we ourselves shape the notions we have of God and of morality. He also understood that we make our language ourselves and thereby give their meaning to the words we use in it, even to the word 'truth'. He seems not to have realised however that by making our language *we also make the rules of logic and mathematics* because they are in the meaning we give to some of its most simple words or signs. That was Tarski's discovery. Popper knew it and recognized in it a methodological principle which he named 'methodological nominalism'. But he seems not to have realised the full extent of its implications. The point is so important that it warrants another

---

110   There seems to be no authorized English translation of that preface. - In the sentence last quoted, Popper must have been referring to Einstein's special theory because that was where Einstein's deviation from Newtonian theory began. My following text presupposes that. For Popper's statement that Newton could not have based his physics on a collection of statements of observation, see *Chapter 2, II*, above.

digression, this time to Tarski's and Popper's theories. In that context I will explain also one of the differences between Popper's view and mine[111].

I

Popper was a staunch supporter of methodological nominalism, there can be no doubt about that. But no philosopher can be great already at the beginning of his endeavours; only the development of his thoughts makes him so. Like Kant himself, Popper too had to work hard for getting rid of old prejudices and for finding his own position of critical rationalism. He gave a vivid account of that struggle in *Unended Quest*[112].

His first acquaintance with the principle of methodological nominalism came only in 1935 through Alfred Tarski[113]. At that time he had already published his *Logik der Forschung* (1934) and had written though not published his treatise *Die beiden Grundprobleme der Erkenntnistheorie*. Only from then on could he begin to realise the implications of Tarski's ideas and to apply them to individual fields of knowledge. Methodological nominalism thus was grafted subsequently on his thoughts. He realised its importance but it remained an alien element in his thoughts to some extent.

Compared to that situation I had the advantage of getting to know Popper's critical rationalism and Tarski's methodological nominalism both at the same time, and of thus being able to regard them from the outset as two inseparable parts of a comprehensive theory. I learned about both of them in 1966 when I read Popper's *Open Society* where he first introduced methodological nominalism in his own works[114]. My own

---

111 Our views differ also in other respects. I will explain those differences in *Appendix 1*.

112 Karl Popper, *Unended Quest* (1979), p. 17–31.

113 Karl Popper, *The Logic of Scientific Discovery*, footnotes to pp. 76, 88, 274.

114 Karl Popper, *The Open Society and its Enemies*. - Critical rationalism in its various aspects is the main subject matter of both volumes of that book. Popper explained methodological nominalism only in Vol. 2, Chapter 11, section II, introducing it as 'a kind of digression' there.

interest in philosophy was only just awakening then and his books roused it more than anything else did. To me, methodological nominalism was an essential part of critical rationalism itself from the outset. The one would have been incomplete without the other, and I owe to Popper's *Open Society* being permitted to see things like that.

As I gradually found my own position these different starting points led to an assessment of methodological nominalism that does not agree with Popper's in all respects. The most obvious difference shows in the way we see Alfred Tarski. Popper summarized his view like this[115]:

> 'In my opinion, *it is not his successful description of a method for defining "true"* which makes Tarski's work philosophically so important, but his *rehabilitation of the correspondence theory of truth*, and the proof that there is no further difficulty lurking here once we have understood the essential need for a semantical metalanguage which is richer than the object language and its syntax.' (Popper's italics).

He emphasized that again in one of his last lectures[116]. Tarski was to him, first, the man who had brought the theory of truth back to reason again.

In my view Tarski's importance lies in method primarily, that is in his method of nominalist definition. Defining the concept 'truth' is but an application of that method, and so is the rehabilitation of the correspondence theory of truth[117]. Both are important of course. Nevertheless they are but applications of a principle that is valid universally and applicable in all sciences.

I believe the name 'methodological nominalism' was Popper's creation. It is also important because it shows that nominalism is a principle of method, at least in science. Nominalism itself had probably been known even before Plato and Aristotle but certainly since the Middle Ages

---

115 Karl Popper, *Philosophical Comments on Tarski's Theory of Truth* in *Objective Knowledge – An Evolutionary Approach*, p. 329.

116 Karl Popper, *A World of Propensities* (1990), p.6.

117 I will discuss those questions in the following section.

through the dispute over the universals. Only Tarski however showed how to apply it consistently as a method by distinguishing between *object language* and *metalanguage*, and by giving non-circular definitions to the terms of logic itself, even to the term 'truth'. Compared to earlier theories, especially to Whitehead and Russell's *Principia Mathematica*, that was an important step because it revealed the underlying principle and permitted applying it methodically.

Popper saw that, but I think he did not see it clearly enough. He deemed possible, at least temporarily and long after Tarski had made him accquainted with the principle (1935), that methodological nominalism might apply only to empirical statements. For non-empirical pure mathematics he not even excluded the possibility of the essentialist method being applicable. He stated that still in his *Open Society* (1945), the very book in which he first introduced methodological nominalism in his own works[118]. At that time Tarski's distinction between different levels of language had already shown that *all* the terms of our language, even the shortest and most important of them, can be given non-circular nominalist definitions. And it had also become apparent thus that *the rules of logic are in the meaning we give to these simple terms.* That illustrates the difficulties Karl Popper had in finding his own position, and it indicates that methodological nominalism remained an alien element in his own world of thought even ten years after Tarski had made him acquainted with it.

II

Tarski's methodological nominalism not only re-established the concept of objective truth valued so highly by Karl Popper[119]. It also showed the

118 Karl Popper, *The Open Society and its Enemies*, vol. 2, Chapter 11, section II (pp. 9–21). His words were ‚I have the empirical sciences in mind, not perhaps pure mathematics' (p. 12).

119 For Tarski's definition of 'truth', see *The Concept of Truth in Formalized Languages* (1930–1932) in *Logic, Semantics, Metamathematics* (1956), 2nd ed. (1983), p. 187–197. –

way to the nominalist understanding of mathematics and logic. It applies to all the different fields of mathematics, but geometry probably is the one best suited for demonstrating that.

The essentialist view assumes the axioms of geometry to be expressions of empirical *truth*. It believes them to originate in observation, for instance in the observation that all points on a circle are at the same distance from its centre. Since the fact itself is indubitable, it believes the circle axiom to be expressing empirical truth. In nominalist understanding however, the word 'circle' is merely a definition which we must therefore read 'from right to left' as we saw in section *II, 3* of the *Introduction*. The word 'circle' is not an answer to the question what a circle *really is*, or what its *essence* is, but an answer to the question what *name* should we give to a line on which all points of are at equal distance from one point. We cannot measure that word by the standard of truth therefore, but only by that of expediency,

Tarski showed that this principle applies also to all other fields of mathematics and to the entire field of logic. The so-called 'rules of logic' are in the meaning we *give* to some of the most simple terms of our language, such as the erms 'is', 'and', 'or', 'only', 'all', 'no', 'not', 'if ... then', 'true', 'false' etc.. By introducing the distinction between *object language* and *metalanguage* he also showed a way of giving all these words formally correct definitions without presupposing the meaning of the *definiendum* in the *definiens*.

We must distinguish the *definition* of the term 'truth' and the *criterion* of the truth of some statement because they have nothing to do with one another. A definition only is about the *meaning* we give to some term in some context, not about that meaning being true or false. - For the purpose of this essay we can define the term 'truth' fairly simply by using Tarski's distinction between object language and meta-language. The definition then consists in saying that a statement belonging to object language (for instance the sentence 'snow is white') meets our definition of 'true' if the same statement is included also in meta-language (in our example therefore if meta-language says that snow is, indeed, white). That way of defining terms is complicated but it avoids circularities. In everyday life it is possible to *explain* the meaning of 'truth' by the words 'correspondence to facts', as Popper did. *Defining* it like that however would only would raise the question of the meaning of the terms 'facts' and 'correspondence'. It would achieve no more thus than shifting to another level the problem of defining 'truth'. – For Popper's discussion of the definition of 'truth' see his *Conjectures and Refutations* (1963), p. 223–227; *Objective Knowledge* (1973), p. 319–340.

For details of this method I must refer to Taski's own papers but I will give one example of a non-circular nominalist definition of a term of logic here. We can define the word 'not' as meaning that any true statement containing that word (eg. 'London is not on the Rhine) will become false by removing it while any false statement containing that word (eg. 'London is not on the Thames') will become true by removing it[120]. Having defined it like that we would be contradicting ourselves by claiming a sentence to be true both with and without the word 'not'. The same applies also to the other simple terms mentioned above. We can give all of them correct nominalist definitions even if some are more complicated.

The ultimate consequence of Tarski's theory is that we can reduce all the rules of logic and of mathematics to only one principle, the principle of not going back on our own actions (*ne venire contra factum proprium*). That in turn is but an implication of Kant's principle of autonomy because 'not going back' stands for a personal decision. And at the level of language (including the language of mathematics) that decision consists in *giving a meaning to words or signs and retaining it in the respective context.* That is the only rule we must observe for being logical (if we intend to be so).

Only this nominalist approach to the theory of logic will permit distinguishing clearly between logic and truth and between mathematics and truth. If we assume the rules of logic and mathematics to be manmade, then we cannot at the same time assume them to be part of nature without violating the meaning of the word 'nature', because using it in contrast to 'manmade' in that context implies that the meanings of the two words are to exclude one another. I believe that principle to be more important even than defining the concept of 'truth'. Only a clear distinction between empirical truth on the one hand and logic and mathematics on the other hand will permit distinguishing clearly between physics and mathematics. And that is the distinction presently violated so often in theoretical physics.

---

120 The definition of 'not' given in the text is incomplete because showing the underlying principle is sufficient in our context. For details, see Alfred Tarski, *The Concept of Truth in Formalized Languages* and Karl Popper, *The Open Society and its Enemies*, vol. 2, Ch. 11 sec. II. (p. 9–21).

The summary of all this is that Popper assigned to the subjective element in the Kantian principle of autonomy its proper place in his theory of scientific discovery, but that he did not always apply it consistently to logic and mathematics. By 'subjective element' I mean the clear commitment to the fact that logic is binding only for those who are willing to be bound. I know that he too was convinced of that. When he first invited me to his house in 1969, one of my questions to him was how it could be that the rules of logic are cogent, yet will be violated so often. I knew no answer to it then and it had worried me in discussions during the students revolts in Germany earlier that year. That is why I can still almost hear the answer he gave me in his brittle voice. He said (in my translation; we were speaking German then).

> 'It depends on whether you want to say the truth or to lie.
> If you want to say the truth you must be logical.'

For some inexplicable reason however he never put that down in writing as clearly as he had said it, although the critical side of his 'critical rationalism' implies just that. It means that rationalism presupposes a decision which can be justified morally but not deductively. He emphasized that most strongly perhaps in his appeal to the intellectual integrity of scientists when he wrote[121]:

> 'Although there is no "rational scientific basis" of ethics, there is
> an ethical basis of science, and of rationalism.'

Nevertheless I think he did not apply methodological nominalism consistently in his writings. He did not give up altogether Kant's notion of an *a priori* valid knowledge because he could not bring himself to accepting that the *only* reason why logic and mathematics are binding lies in those willing to be bound.

---

121 *The Open Society and its Enemies*, vol. 2, p. 238.

(*1*) In Popper's time the problem of distinguishing between mathematics and truth arose in geometry, and its greatest difficulty consisted in finding the correct approach to understanding geometrical axioms. Anyone reading Henri Poincaré's *La Science et l'Hypothèse* (1902) and seeing how that open-minded scientist literally wrestled with the problem there will probably agree with me on that. When Popper started working on that problem he must still have stood strongly under the impression of Einstein's *Geometrie und Erfahrung* (1921). The passages quoted in *Chapter 2* from that paper leave no doubt that its influence could not have beneficial to him for making up his mind whether he should hold axioms to be true or to be valid, and if so, why[122]. That must have been why his position on the issue was ambiguous from the start. In his *Logik der Forschung* (1934) he explicitly refused to establish the truth or validity of the axioms of geometry by declaring them to be 'immediately certain', or 'intuitively certain', or 'self-evident'. But he also left open *explicitly* whether axioms were to be regarded as conventions or as empirical or scientific hypotheses, deeming *both* of these interpretations to be 'admissible'[123]. Thus he also left open whether the axioms of geometry can lay claim to empirical truth or are subject to the standard of expediency. He did not say so explicitily, but neither did he take sides on the issue[124].

Popper was a child of his times of course. He could detect the superstition in which he had grown up, and could bemock the mental fetters it put on epistemology. But he could not free himself of them altogether[125]. At the time of writing his *Logik der Forschung* he still believed logic and

---

122   He mentioned Einstein's *Geometrie und Erfahrung* in *Die beiden Grundprobleme der Erkenntnistheorie*, of which his *Logik der Forschung* was originally intended to be a part.

123   Popper, *The Logic of Scientific Discovery* p. 72

124   My own view comes very close at this point to that held by Henri Poincaré in *La science et l'hypothèse* (1902), Chapter 3, subtitle '*De la nature des Axiomes*'. Poincaré said there (in my translation) 'In other words, the axioms of geometry (I am not speaking of arithmetic) are but definitions in disguise'. He also said that the question of Euclidean geometry being true had no meaning whatsoever ('aucun sens') because the standard for gauging it was that of expedience ('commodité').

125   The text refers to the motto from Lessing's *Nathan the Wise* quoted at the beginning of this chapter.

mathematics to be *systems* and was deeply impressed by Einstein's stating that the general theory of relativity would be refuted if there were no deflection of light in the gravitational field of the sun[126]. In that book he also deemed possible

> 'for the primitive concepts of an axiomatic system such as geometry to be correlated with, or interpreted by, the concepts of another system, e.g. physics'[127].

That was replacing Kant's assumption of Newtonian physics being true *a priori* by Einstein's assumption that extending it by another axiom would convert the non-empirical axiomatic system of Euclidean geometry into the empirical system of physics .

In his treatise *Die beiden Grundprobleme der Erkenntnistheorie*, written at the same time as his *Logik der Forschung* (1930–1933) but published only in 1979, the ambiguity of Popper's attitude at that period showed even more clearly. That may have been why he delayed publishing it for so long. In section (30) of that book he discussed at some length the conventionalist and the empiricist interpretation of axiomatic systems without clearly committing himself to either of them. He even favoured the empiricist interpretation there, arguing on those grounds against the criticism Poincaré had brought forward against it[128]. That was why he also deemed possible that Einstein's axiomatic understanding of physics, as related in *Chapter 3* of this essay, could be correct. Seeing things like that was natural in a way because interpreting geometry as an empirical science will almost inevitably lead to an axiomatic understanding of physics, especially when linked with believing in 'systems'. To my mind however, understanding physics as an axiomatic system stands in open conflict with conceiving it as an empirical science as Popper's criterion of falsifiability correctly presupposes.

---

126 Popper gave a vivid account of that in *Unended Quest*, p. 37, 38.

127 *The Logic of Scientific Discovery* p. 75.

128 In *La science et l'hypothèse*, Part II, chapter III, Poincaré had argued strongly against the view that the axioms of geometry can claim to 'truth' in any sense of the word. In his view they were 'définitions déguisées' (p. 55), in other words they were mere conventions.

The haziness in Popper's way of treating the axioms of geometry influenced also his theory of logic. He had given the right answer in his above-mentioned reply to my question why the rules of logic are cogent yet can be violated. But he never quite overcame the ambiguity of his own views, not even in his later writings. That is why I think it must have been his belief in logic and mathematics being *systems* that stood in his way. He never really accounted clearly and unambiguously for the relevance of logic and mathematics in science. In his *Open Society* (1945) he still deemed possible to gauge statements of pure mathematics by the standard of *truth*[129]. Later he described logic 'as the theory of deduction or derivability or whatever it may be called'. And he assumed the function of valid conclusions to be that they will 'transmit to the conclusion the truth of the premises'. In his paper *Why are the Calculi of Logic and Mathematics Applicable to Reality?* (1946) he *explicitly* left unanswered the question posed by that title[130]. And he never, to my knowledge, discussed it again in any of his later writings.

Popper's wording sometimes even indicates that he seems to have regarded the rules of logic as empirical laws. He described a valid rule of logical inference as being

> 'useful because no counter-example can be found, that is, because we can rely on it as a rule of procedure taking us from true descriptions of fact to true descriptions of fact'[131].

---

129 Karl Popper, *The Open Society and its Enemies* (1945), vol. 2, p. 15f. - He wrote there: 'Everybody who "understands" an idea, or a point of view, or an arithmetical method, for instance multiplication, in the sense that he has "got the feel of it", might be said to understand that thing intuitively; and there are countless intellectual experiences of that kind. But I would insist, on the other hand, that these experiences, important as they may be for our scientific endeavours, can never serve to establish the *truth* of any idea or theory, however strongly somebody may feel, intuitively, that it must be true, or that it is "self-evident".' (my italics). – The text shows that he regarded multiplying as a *method*, and wanted to measure understanding it by the standard of *truth*. In the note to that text (p. 291, note 43), he treated the sentence '2 + 2 = 4' as a *true* statement.

130 *Conjectures and Refutations*, p. 201, 213, 214.

131 *Conjectures and Refutations,* top of p. 205.

By referring to 'counter-examples' he was applying his criterion of falsifiability to the rules of logic themselves, placing them thus on a level with empirical laws of nature. Consequently he would have had to apply the standard of truth for assessing them (instead of the standard of expediency which I believe to apply). He never said so explicitly but the possibility must have been present in his mind because it occasionally turns up in his writings. On the one hand he said:

> 'Rules of inference are procedural rules or rules of performance, so that they cannot "apply" in the sense of "fit" but only in the sense of being observed. Thus a world in which they do not apply would not be an illogical world, but a world populated by illogical men.'[132]

On the other hand he argued further down in the same essay that

> 'the rules of logic are laws of certain descriptive languages – of the use of words and especially of sentences.'[133]

That presupposed their validity being independent of the individual applying them.

At no point of his writings that I know of did Popper say that we not only decide for or against *observing* the rules of logic, but that we actually *create* them. In the first passage just quoted, the image he used of 'illogical men' populating our world even tells against his having seen things like that. He was using the word 'logic' in the strictest sense in that context. But in that strictest sense there are no illogical people but only illogical statements. Even the most rational of people may be irrational at times, and even the most irrational of people can sometimes think and act rationally. People react to situations. That is why they can react this way today and that way tomorrow. For the same reason they can decide for or

---

132 *Conjectures and Refutations,* top of p. 205.
133 *Conjectures and Refutations* p. 207.

against logic. And they do so not just once in their lives and then forever, but whenever they use language as a means of communication. Their statements can be logical this moment and illogical the next because at every moment they themselves create their language and thus the meaning of their words. That is my view, and I think seeing things like that is a consequence of Popper's own theories, especially of that of the 'Three Worlds' and of his belief that 'All Life is Problem Solving'[134]. Popper however seems to have regarded the rules of logic as being a predetermined, ready-made system of standards.

(2) In a sense Popper thus upheld a last relic of Kantian *a priori* at this point. Despite his justified criticism of that part of Kantian theory he still considered logic, mathematics and geometry to be knowledge valid *a priori,* though in a modified sense. I think he may have been reasoning along the following lines. We can decide for or against *observing* the rules of mathematics and geometry. That is where Kant's principle of autonomy comes in. Those rules however belong to a *system* of rules (as he must have believed). Changing only one of them would therefore knock over the entire system. For that reason they could be accepted or rejected only as a whole.

This interpretation would fit in quite well also with Popper's theory of the 'Three Worlds' mentioned above. Logic, mathematics and geometry are products of the human mind. They belong to his World III. He had shown convincingly that though invented by man, their existence from then on is independent of man. Like the stone of Rosette they would continue to exist even after no human being lived anymore. His next inference might then have been that the systems of logic, mathematics and geometry not only *exist* independent of man but that the rules of which they are composed must also be *valid* independent of man.

My interpretation is purely conjectural of course, but it would also explain why Popper was more interested Tarski's rehabilitation of the correspondence theory of truth than in in the methodological principle

---

134 For Popper's theory of the ,Three Worlds', see *Objective Knowledge* (1973), Chapters 2–4 and *The Open Universe* (1982), p. 116–130.

revealed by his theory. In the view just described, logic and mathematics could not have been subject to negotiation from the outset. If Popper assumed them to be predetermined, ready-made systems of standards which can be accepted or rejected only as a whole, then there would have been no need to apply to them any method at all, not even that of methodological nominalism. They required no method according to that view. Rather, they *were* method. In any case they were valid objectively and, so he must have thought, self-evident even without any method at all.

Popper learned only about in 1960 from a discussion with Imre Lakatos (1922–1974) that logic is no system at all but *open in principle*, and that even mathematics is no system but also open in principle. He mentioned that in a paper read in 1965 and published only posthumously, but he never mentioned the consequences arising from that view[135]. I cannot say therefore whether he ever realised that we not only decide for or against the rules of logic and mathematics but that we actually *create* them ourselves by giving their meaning to the words we use. I could find no statement to that effect in his works and we never got down to such issues in our discussions[136]. Some passages in his works however make it seem unlikely that he could have seen things like that. That is why I think the impression which Einstein's belief in the necessity of an 'exact science' had made on him in his earlier years must still have been too strong. Yet the rich development of mathematics since the times of Euclid leaves no doubt that the rules of logic and mathematics are products of the human mind. Man created them. Popper's own critical rationalism should have been even more critical at that point. That seems to be the deepest reason of our disagreement.

---

135 Popper, *Beyond the Search for Invariants*, in *The World of Parmenides*, p. 146 ff., 152 f.

136 Our discussions were on a problem arising in legal theory. I believe it to be analogous to the one raised in the text, and we strongly disagreed on it. Popper had discussed the relation of norms and facts in his *Open Society* (vol. 1 ch. 5, esp. Note 5; and addenda vol. 2, pp. 369–396) and there proposed to assume a 'critical dualism'. In my *Recht und Rationalität* (1984) I criticized that as being essentialist in the approach, and took the view that according to Kant's principle of autonomy not only must every man 'make his god himself', as Kant himself had said, but similarly every lawyer also 'must make his legislator himself'. I reported on that controversy in my paper *The Problem of Objectivity in Law and Ethics*, included in *Popper's Open Society after 50 Years* (1999), pp. 111–127.

(3) Popper must have assumed between a system and its objectivity a connection that does not exist. I think he must have confused the question of the *objectivity* of logic and mathematics and that of their *validity*.

Objectivism is extremely important in Popper's theory but it does not depend on anything being valid independent of man, not even logic and mathematics. In usual understanding the epithet 'objective' stands for an assessment by criteria that are independent of the respective observer and open to inter-subjective criticism in Popper's sense if possible. In that sense, logic and mathematics need not be ready-made *systems* of rules for deserving the epithet 'objective'. If it is possible to observe and to criticize whether the meaning of a word or a symbol remains the same in some context, then that will be sufficient for safeguarding the objectivity of a logic and mathematics. At any rate it is the best we can get.

New rules can also be developed as it happened so often in the history of mathematics, for instance in the development of analytical geometry, or of the integral calculus, or of non-Euclidean geometries. They can even be developed in logic, for instance by adopting Tarski's rule that

> 'whenever, in a sentence, we wish to say something about a certain thing, we have to use, in this sentence, not the thing itself, but its name, or designation'[137]

The rules we need depend on the problem we want to solve, in logic as well as in mathematics. The example of the German mathematician Carl Friedrich Gauss (1777–1855) illustrates that. The government of Hannover had commissioned him with a complete survey of the kingdom. Its territory included plane areas too large for neglecting the curvature of the earth's surface. He therefore had to determine distances on that surface without being able to measure them in straight lines by optical means. Others might have travelled those distances for getting precise measurements in that situation. Gauss integrated the curvature of the earth's surface in his system of coordinates and eventually solved the problem by

---

137 Alfred Tarski, *Introduction to Logic* (1965), p. 58–60.

inventing spherical geometry[138]. The relationship between system and objectivity exists only seemingly therefore, and only for those believing in ready-made 'systems' as Popper must have done. He believed in the possibility of an *Open Society* but he could bring himself only gradually to believe also in an *Open Science*. But if we can reduce logic and mathematics to the single rule that words and symbols may have only one meaning within one context, then the fact that compliance with that rule is open to inter-subjective criticism will safeguard their objectivity.

III

Let us now be back with Einstein again. We saw that he had ventured to ascribe the discoveries made by science to free inventions of the human mind. But though following Kant up to that point, he still could not free himself from the notion that knowledge, at least scientific knowledge, must be absolutely certain and indisputable. He rejected Kant's notion of a knowledge valid *a priori*. But he believed that the absolute certainty of science which Kant had sought and which he too considered indispensable, could be found in the strict rules of logic and geometry which in his conception nature itself had given. So he proposed to extend the non-empirical science of geometry by one more axiom and thus to convert it into the empirical science of physics. That, so he believed, would endow also physics with the absolute certainty of geometrical statements. He did not follow to its end the path he had taken, and that was to be his undoing.

(*1*) Einstein ended the search for a deductive justification of axioms but he did not end the search for the axioms themselves, not even in physics. Although rejecting the idea of any kind of knowledge whatsoever being valid *a priori*[139], he nevertheless stood unsveringly by his belief in the axiomatic

---

138 C.v.Mettenheim, *Popper versus Einstein*, p. 81f).

139 See the excerpts from Einstein's papers on *Geometry and Experience* and *On the*

method. And he continued to search for those last, reliably secured truths on which science was to build. Euclid's geometry he described as the 'mental miracle of a logical system' and as an 'admirable achievement of reason'. So he had realised that mathematics 'is a product of human thought independent of all experience'. Yet he also believed it could be a source of *empirical* knowledge. 'Pure mathematical construction', he asserted, 'enables us to discover the concepts and the laws connecting them, which give us the key to the understanding of the phenomena of nature.' And he contended that 'the truly creative principle resides … in mathematics'.

To my mind there is no denying that Einstein's thinking was dark and contradictory at this point. He did not say what criterion ultimately was to decide on the demarcation between empirical science and non-empirical science. On the one hand, mathematics and geometry were to him 'product(s) of human thought … independent of experience', which put them into *non*-empirical science. On the other hand he expected them to 'give us the key to the understanding of the phenomena of nature'. That would require that they include, at least among other things, also statements about phenomena of nature, and these would therefore have to be empirical statements. Both assumptions cannot be true at the same time however, because if mathematics were to be empirical and non-empirical, then that would violate the principle of non-contradiction located in the definition of 'non'.

I hope the quotations in *Chapter 2, III* show that the contradiction just demonstrated does not originate in my interpretation of Einstein's texts but directly in his words. Anyone doubting that should re-examine those quotations under this aspect. They indicate that to Einstein mathematics was a creation of the human mind and *at the same time* a source of empirical knowledge. That was why he deemed possible, indeed necessary, to search for the fundamental axioms of empirical knowledge. In his view the scientist no longer was the craftsman helping to build the great edifice of science to a foreign plan. He could even believe to be himself the

---

Method of Theoretical Physics quoted in *Chapter 2, III, 1, 2*. The quotations following in the main text are from those excerpts.

architect. The edifice however continued to exist and still was to stand on secure foundations. And what other foundations of science could have been more secure than the rules of logic and geometry?

Einstein's thoughts unfortunately were half-hearted at this point. He did not pursue to its end the courageous approach which Kant had shown and he himself had taken, but timidly shirked the consequences it foreboded. The notion that scientific knowledge might really be a creation of the human mind, and *never* be reliably secure, would have meant that we bear a personal responsibility for what we deem to be the truth. That appeared to him far too daring to be actually upheld. So he continued pursuing the axiomatic geometrical approach, ostensively shifting responsibility away from the scientist on science itself thereby. Following that path and never leaving it to the end of his life led him into those conceptual contradictions and logical circles from which he was ultimately not to escape.

(2) The breach in Einstein's theory of knowledge may become clearer by directly comparing his ideas to Karl Popper's. In the history of philosophy Popper was the first to go to its end the road Kant had opened and Einstein had first taken but then abandoned.

Popper always recognized that Einstein had given him important inspirations for his *Logic of Scientific Discovery*. He often quoted and generalized Einstein's famous answer to the question how it could be that 'mathematics, although a product of human thought and independent of all experience, will fit the objects of reality so well'. Einstein's answer had been that

> 'As far as the laws of *mathematics* refer to reality, they are not certain; and as far as they are certain, they do not refer to reality.'[140] (My italics).

---

140 Albert Einstein, *Geometry and Experienc*. See *Chapter 2, III, 1*, above.

Popper turned that into the seemingly matching but on scrutiny far more radical and much clearer statement that

> 'insofar as the *theorems of science* refer to reality, they must be falsifiable, and insofar as they are not falsifiable, they do not refer to reality.'[141] (My italics).

It was more than a mere generalization when he replaced the word 'mathematics' of Einstein's statement by the word 'science' in his own. He thereby introduced the distinction between empirical and non-empirical science which Einstein had failed to see. In Einstein's statement mathematics could refer among other things *also* to reality. It could be empirical or non-empirical. Popper's statement shows that he still believed science to consist of 'theorems' at that time. Nevertheless his criterion of falsifiability divides knowledge in two categories, empirical and non-empirical, and thereby separates them sharply. Empirical knowledge must be falsifiable; and non-empirical knowledge, including mathematics and logic, does not refer to reality. Popper thus avoided from the outset the self-contradictions of Einstein's thought.

Falsifiability was not only the criterion of empirical science in Popper's theory however. He even raised it to the rank of a *postulate*. That supplement, seemingly atmospheric only, was in fact a radical innovation. It was so radical that its importance still has not been understood sufficiently. Popper slightly modified his opinion on the demarcation between science and non-science in the course of his life, even though admitting that only reluctantly. In this important point however he never weakened it but continuously strengthened the emphasis he put on it. In his theory improbability and refutability of a scientific theory no longer were expressions of human shortcomings, therefore of personal weaknesses bashfully concealed. Instead, he recognized in them the highest virtues of *empirical* science.

A scientific theory leaving open surface to empirical criticism is not weak for that reason in Popper's theory. On the contrary, the more it

---

141   Karl Popper, *The Logic of Scintific Discovery*, p. 314.

exposes itself the stronger it will be – provided it survives the attacks. Newtonian theory for example takes its informational content from its mathematical precision which shows an open flank to empirical criticism at any time. A stone's throw would refute it if persistently it defied the laws of gravitation. Yet it withstood all empirical criticism so far. That makes out its strength.

Conversely, a theory not open to empirical criticism is suspicious as a *scientific* theory for that very reason in Popper's theory. He later conceded that theories of that type need not be worthless from the outset[142]. Darwin's theory of evolution is his most famous example. It is not open to criticism on empirical grounds because no experiment could refute it. Nevertheless, as an interpreting theory or, as Popper put it, as a 'program of metaphysical research' it is an invaluable contribution to our understanding of evolution and to the progress of science. We will see in *Chapter 9, VI, 3* that related principles apply also to some other theories.

Popper thus was the first philosopher to draw from David Hume's argument against inductivism the radical conclusion that in the field of empirical laws of nature there can be no provable knowledge at all. Scientific knowledge of those laws must forever remain hypothetical.

(3) We will often be back with Popper's theory of science in this essay[143]. Comparing it with Einstein's ideas ought to have shown already though that Einstein and Popper were very close to each other in their approaches to the theory of knowledge in the natural sciences, but that they differed in one crucial respect.

For the origin of a scientific theory, that is to say for its first invention, both did not rely on a constructively rational foundation of the theory but referred to the intuition of the scientist, recognizing thereby that intuition cannot be rationally *justified* in any way whatsoever. Up to that point they both followed Kant's 'Copernican revolution of epistemology' by trusting in the creative power of the scientist and by regarding knowledge of nature

---

142 See Popper, *Unended Quest*, p. 16 8f.

143 In *Appendix 1* I have tried to give a comprehensive summary of the questions on which I disagree with him.

not as something conferred on man from outside but arising from his own autonomous and creative mental power. That was the Kantian anthropocentric element of their approach in which understanding nature is no passive process but an active performance, coming from man himself.

Einstein took an important step beyond Kant by denying the existence of any knowledge valid *a priori* and by consistently doing away with any attempt at constructively justifying scientific theories[144]. Instead, he relied only on subsequent examination of a theory with regard to its being suited for solving a problem (heuristics) and its conforming to experience (empiricism). In the history of epistemology he was the first to venture taking that step. That must assure him lasting fame.

His thoughts became vague and contradictory however, when in spite of that approach he still regarded mathematics as a possible source of empirical knowledge and as the 'truly creative principle' of science. He in fact revoked the principle autonomy thereby and replaced it by his own belief in the empirical truth of mathematics. That was to be his *a priori* valid knowledge even if he did not see that.

Popper accepted Einstein's proposal to recognize the 'free creation of the human mind' as a legitimate path to knowledge. He, too, did away with all constructive foundation of scientific theories and referred only to subsequent examination with regard to their being suited to solving problems (plausibility) and to their conforming to experiments (falsifiability). In that heuristic and at the same time empiricist approach both philosophers agree.

From then on however their ways separate, at least in theory of knowledge. Einstein fell back on mathematics and continued to search in empirical science for ultimate, reliable truths which he hoped to find in its basic axioms. Popper, by contrast, insisted on asserting that in the field of empirical science there can be no certain knowledge at all. In his theory *all empirical knowledge of the laws of nature is only hypothetical, and will forever remain so.* Our entire knowledge in that field consists of mere con-

---

144 For details see the quotations from Einstein's papers on *Geometry and Experience* and *On the Method of Theoretical Physics* in *Chapter 2, III, 1, 2.*

jectures, distinguished from erroneous theories only by the fact that no observation ever refuted them[145].

Popper's theory of falsifiability thus stands for the most radical scepticism ever held by anyone before him in the theory of knowledge. And being more radical it also is more consistent than any previous theory. It is strictly objectivist, containing neither subjectivist nor nihilistic elements.

Subjectivism it avoids by replacing the certainty of axioms or of experience through the criterion of *intersubjective testability*[146]. Subjective influences are unavoidable even in science. But we can eliminate their effects by demanding that empirical scientists must state their theories and observations in a way that enables others to repeat them and to criticize them. Being open to criticism by others saeguards the objectivity of scientific empirical knowledge in Popper's theory. It also avoids nihilism because instead of pretending to a certainty of knowledge not existing in reality, it appeals to the moral responsibility of scientists for accepting empirical criticism and thus furthering the progress of science.

Einstein's supposedly third way to cognition, by contrast, which was to avoid the logical problems of deductivism and inductivism by introducing a new axiom, ended in fact by leaving unresolved all the problems of empirical knowledge. According to his theory empirical knowledge in the field of the natural sciences includes, at least among other things, also knowing the laws that govern the phenomena of nature. So he did not avoid the question of how such knowledge is possible. But neither did he answer it. He simply evaded it by shifting the goal of scientific research itself, assuming the aim of science to lie not in expanding *empirical* knowledge but in discovering *mathematical* truths of nature.

People living with contradictions may be highly sensitive however. They may own the gift of realizing contradictory intellectual currents and of making visible positive trends. Einstein owned that gift. He felt the

---

145 That was Popper's key statement in his *Logik der Forschung* (1934 - The Logic of Scientific Discovery).

146 For 'intersubjective testability', see Karl Popper, *Logik der Forschung* (The Logic of Scientific Discovery,) section 8.

spirit of Enlightenment which had brought about all those wonderful discoveries of the 19th Century. By assuming that nature will *per se* behave mathematically and by transferring to empirical science the axiomatic method of Euclidean geometry he tacitly adopted, though without noticing that, Kant's problematic assumption of an *a priori* valid empirical knowledge which he explicitly rejected[147]. To that extent he was contradicting himself. But by claiming the axiomatic basis of physics to consist in 'free creations of the human mind' and by deeming possible that 'pure thought is competent to comprehend the real, as the ancients dreamed', he also took up Kant's truly revolutionary idea that man himself invents the laws of nature and examines them only subsequently at the hands of experience.

In spite of the contradictions in Einstein's theory it therefore remains a historic achievement that he demonstrated by his heuristic method how speculative reasoning combined with *ex post* control by experience can bring about the advancement of knowledge. He was the first to point to the basic difference between the old (uncritical) deductivist approach and the new (critical) deductivist approach. And he thus gave Popper the decisive inspiration for his modern theory of scientific knowledge. The task of science must now be to apply systematically that theory to theoretical physics in order to see what remains after that.

---

147 See *Chapter 2, III, 1.*

CHAPTER 6:

ON THE AIMS OF SCIENTIFIC RESEARCH

Impacts on scientific research of different approaches to epistemology; I. Fictitiously applying the axiomatic approach: collecting theories; reducing the number of axioms. II. Psychological conditions of the 'axiomatist': science infallible in principle; errors personal shortcomings; example for discovery: Hertz' of photoelectric effect; for axiomatist, discoveries are setback; Einstein's attitude on discoveries, examples: redshift; Hubble effect; correction of GTR by Hubble constant; Sagnac's experiment ignored. III. Interpretation of experiments; priority of theory, examples: Galvani, Volta; discovery of nuclear fission; discovery is theory, not observation (Popper): Copernicus' theory; rescuing theories by interpretation; deciding by theory or by experiment, examples: nuclear fission and e = mc²; time dilation and change of method; IV. Axiomatic theory obstructs progress of science by hampering discoveries.

The problems of epistemology discussed in previous chapters raise a new question. Are the different approaches to those problems relevant for philosophers only or will they influence physicists also in their research work? I think there must be some kind of an impact. Few human beings will permanently tolerate an open conflict between their actions and their convictions. Most will wish to create a kind of pre-established harmony in that field by letting them converge somehow. Not all, of course, will reach that happy state by aligning their actions to their convictions. Many will prefer adapting convictions to circumstances. From the point of view of convenience that way clearly is preferable.

Einstein however was of that rare species who would act from conviction even when circumstances were unfavourable. In 1918 he had said in his speech on Max Planck's 60th birthday that 'the highest task of the physicist is thus to find those general elementary laws from which, by pure deduction, the worldview can be gained. No logical path will lead to these laws but only intuition based on an emphatetic understanding of experience.'[148]

By that he stood. He was convinced that 'those most general elementary laws' were the supreme goal of all theory[149]. So he searched for those most general elementary laws. In almost four decades of his life he dreamed of a general field theory and of an even more general 'world formula' that was to include in an all-in-one-solution all the basic principles of physics. He not only invested most of his energy in his search for that formula but even sacrificed for it his reputation as a scientist. Owing to his early achievements, his theory of relativity and his involvement in the beginning of quantum theory, he continued to be the most admired physicist of his time. For his later work however he hardly found more than indulgent smiles among his contemporary scientists[150]. In his theory of relativity he had denied the absolute meaning of 'time'. It sometimes seemed as if time was taking its revenge now by simply passing over him. As a scientist he was buried alive more than two decades before the end of his life.

His perseverance and courage in pursuing his quest tell of his strong character and of the sincerity of his convictions. They deserve admiration for that, but they also raise a question which must interest not only psychologists but also philosophers of science. We must ask whether Einstein had properly selected the goal of his research.

---

148   My translation from Einstein, *Prinzipien der Forschung*, address at the occasion of Max Planck's 60th anniversary, in *Mein Weltbild*, p. 175 ff., 178.

149   See also *Zur Methodik der theoretischen Physik* in *Mein Weltbild*, ed. by Carl Seelig (1991) p. 185–195 (On the Method of Theoretical Physics, Philosophy of Science, vol. 1 [1934], p. 163), quoted in *Chapter 2, III, 2*, above.

150   See Albrecht Fölsing, *Albert Einstein - Eine Biographie*, p. 790 ff.

That question concerns not Einstein alone but the whole of theoretical physics in the 20th Century. We might ask it similarly of almost any other theoretical physicist of that period. Before trying to answer it, and following the example of Einstein's heuristic method, we must investigate in a first step only the plausibility of his approach. That seems to be a necessary precaution against wasting time and energy in a search that might prove futile if it should turn out that he had wrongly determined his goal.

We must consider what the world of physical science would look like if Einstein's approach were correct. To that end I propose that we take him at his word. As he did in his famous thought experiments, and strictly by the standards of his own papers, we will try to put ourselves mentally in the situation of a scientist believing his most important task to consist in making the 'irreducible basic elements (of physical science) as simple and as few as possible'. This fictitious scientist is to believe unswervingly that it is possible in principle 'to find those general elementary laws from which, by pure deduction, the worldview can be gained'. At the same time however he must be determined that this reduction shall not 'surrender the adequate representation of a single datum of experience'. Experience will be the ultimate judge in his science at all events.

Our question in this chapter will be what methods a scientist would have to apply for putting into practice Einstein's principles. I will call him the 'axiomatist' from here on, and I think he would have to proceed like this.

(*1*) The axiomatist is unwilling to 'surrender the adequate representation of a single datum of experience'. He must be aware of all human experience previously made. For achieving that goal he would first have to collect the empirical knowledge available in his time, and to arrange it in a way that would make it manageable.

We learned from Karl Popper's refutation of inductivism, however, that observation itself always is the result of choice. I remind of his proposal 'Observe!' related in *Chapter 2, II*. Popper had deliberately chosen it absurd in order to show that no observation is possible without the observer previously deciding on what to observe. Similarly, no collection of observations is possible without previous choice either. That was Einstein's view too by the way, at least according to a report by Heisenberg[151]. It applies even if we really did attempt to collect *all* observations man ever made, which incidentally would take up all the time that ever was for making them and would therefore be palpably absurd as a goal of science. It is as impractical as drawing a map on a scale of 1 : 1 which also cannot possibly succeed.

No axiomatist can begin by simply collecting pure observations and arranging them only afterwards. Even in collecting them he must already make a decision. And if that decision is to make any sense, then he must make it in the light of some criterion. In the case of science only a scientific theory could provide that criterion. The axiomatist must look for information appearing relevant in the light of the theory of his choice. Only then can he arrange it in order to make it manageable.

Our axiomatist cannot begin his quest by starting from observations therefore. In the first step of his efforts he must select suitable theories or frame them himself. Only then, in a second step, can he collect the observations relating to those theories. In doing so, however, he must remember that the observations must be compatible with the respective theories. For, as Karl Popper showed, a theory breaks down if it clashes with reliable observation. The efforts of our axiomatist must begin by collecting all the scientific theories that were never refuted by experiment.

(2) Despite serious difficulties the work of the axiomatist up to that point seems closely to resemble that of any other scientist determined to deal comprehensively with the field of his science. But any ordinary scientist would already have reached his goal with that. If he had really collected

---

151 In *Erinnerungen an die Entwicklung der Atomphysik in den letzten 50 Jahren* (now in ‚Deutsche und Jüdische Physik‘, p. 187, 194) Heisenberg referred to a personal communication by Einstein with that content.

all scientific theories that never were refuted by experiment, then he would have achieved more at any rate than any scientist had achieved before him, and probably more than anyone after him ever will. The axiomatist having achieved all that however, must now turn to the higher goal he has set himself, the goal to

> 'make the irreducible basic elements as simple
> and as few as possible.'[152]

After the first step of his efforts consisting in collecting or stating suitable theories, and the second step consisting in collecting the observations relevant in the light of those theories, the axiomatist now has to take a third step. He must reduce his theories to 'those general elementary laws from which, by pure deduction, the worldview can be gained'. His task is far more difficult than that of any ordinary scientist, his aim much higher.

II

The objectives of research of our axiomatist differ significantly from those of an empirical scientist, and the difference is relevant not only in theory.

The axiomatic approach has a tendency to generate mental conditions which almost inevitably must affect research itself. In axiomatic understanding science is infallible in principle because it is assumed to be based on secure knowledge and to proceed from there by logical inference. If mistakes occur in spite of that, then they can only result from lack of diligence or from mental deficiency, at any rate from some personal shortcoming of the scientist. Anyone believing in the axiomatic approach and endeavoring a carreer as a scientist will thus be tempted to present himself

---

152 *Zur Methodik der theoretischen Physik* in Mein Weltbild, ed. by Carl Seelig (1991) p. 185–195. (On the Method of Theoretical Physics, Philosophy of Science, vol. 1 [1934], p. 163 (quoted in *Chapter 1, II, 2*).

as being infallible, and to cover up failures. Instead of encouraging all kinds of rational criticism because it advances the progress of science, he will be inclined to take personal offence at it. I believe that stifling mechanism at the level of psychology to be the origin of the present crisis of theoretical physics, and will try to illustrate that by examples.

(1) In 1886 Heinrich Hertz irradiated a spark gap with invisible ultraviolet light and observed the brightness of the sparks increasing under the influence of irradiation. The observation did not confirm his expectations at first because he had not intentionally irradiated that spark gap. In his overstuffed laboratory it had only accidentally been exposed to ultraviolet light having nothing to do with the experiment he was carrying out at the time. He did not know that ultraviolet light could have an effect on electrical sparks, and did not anticipate that irradiation would influence them. In fact, he not even realised at first that the effect had been caused by irradiation but first had to search for that cause[153]. He took that as a challenge to his imagination however, and in the further course of events it made him a happy man because he reached the supreme goal an experimental physicist can hope to attain in his discipline. He discovered a previously unknown physical effect, the photoelectric effect[154].

Even discoveries caused by chance can thus be blissful events for an experimentalist. An experimental physicist discovering a hitherto unknown physical effect, even if more or less by accident, has his dreams come true and will often have reached the goal of his desires. He has cheated nature out of one of its well-kept secrets and has learned something new for himself. He can be happy with that because it will add to his own knowledge and to that of mankind. His worldview will change however. He may even realise that he had previously misunderstood something. Sometimes the discovery will strike him like scales falling from the eyes. But since

---

153 I am following Fölsing's account in *Albert Einstein – Eine Biographie*, p. 164, and *Heinrich Hertz – Eine Biographie*, p. 283 ff.

154 Wilhelm Hallwachs (1859–1922) achieved the systematic exploration of the photoelectric effect, but according to Fölsing's reports Hertz was the first to discover it.

it extends his own knowledge and that of humanity it will give him joy, perhaps even amounting to triumph.

The axiomatist, by contrast, must feel quite differently about situations of that sort. His aim is not discovering hitherto unknown physical effects, it is far more ambitious. He wants to make the 'irreducible basic elements' of his science 'as simple and as few as possible'. At the end of this reduction he hopes will stand the single world-formula comprising all physical knowledge about this world. In pursuit of that noble objective he must know and take into account all physical effects because he is unwilling to 'surrender the adequate representation of a single datum of experience'. He must be the mastermind therefore, uniting in himself all empirical physical knowledge of mankind. He starts from theories consistent with experience and tries to find the common features underlying them. But his ultimate goal of research is not in discovering individual physical effects. It is in finding

> 'those general elementary laws from which, by pure deduction, the worldview can be gained'.

(2) New discoveries of physical effects can hardly be welcome to the axiomatist on that quest. If they really were unknown, then even he himself could not have known them. And if he had not known them, then he could not have made allowance for them in his theories or in his formulae.

He will gladly accept such discoveries if they agree with his own theories. In that case he can regard them as welcome confirmations, but only in that case. If a discovery gets in conflict with his own approach however, if it is really new, perhaps even incompatible with the thoughts he had been pursuing, then he should be pleased in his capacity of student of nature, but in his capacity of axiomatist it can cause him no joy. In fact, he would need an extraordinary share of human generosity if it did not appear to him actually disturbing. The more unforeseen it is the more it must worry him. Einstein saw that clearly when in 1934, long after beginning his quest for the unified theory, he almost resignedly wrote in a personal letter to Max von Laue:

'The great discoveries give me only little joy because they do not seem to me, at present, to make understanding the foundation any easier'[155].

How indeed could he have been pleased if each of those great discoveries showed him that his own knowledge must either have been incomplete or that at least one of the theories on which he had relied must be mistaken? Must he not have seen personal defeat in every new discovery? Must he not almost inevitably have been disappointed by it? Did not each of those discoveries confirm Michael Faraday's observation that

'experimental physics ... is a great destroyer of preconceived ideas?' [156]

On the other hand, from what base was he to start if he wanted to continue searching for 'those general elementary laws'? Was it not essential to start from theories agreeing with experience? Did he not have to start from 'preconceived ideas' therefore? Yet how could he consider new discoveries if they were unknown when he began his search?

Clearly, our axiomatist cannot solve this problem any more than Einstein could solve it in his time. Every new discovery will tell him that he had been starting from wrong assumptions. At least that is what it ought to tell him. In that case however it will threaten to wreck his work. If he fails to reconcile his own theories with the new discovery, then it will even endanger the aim of his research. For apart from refuting one of his premises it also means that the system of axioms for which he is searching may have to become more complex than it had previously been. It might even require introducing an additional axiom. His goal of *reducing* the number of axioms and of discovering the one fundamental unified 'world formula' would then be still further removed.

---

155 My translation from Fölsing, *Albert Einstein - Eine Biographie* (p. 780). It seems likely to me that Einstein was referring to the Hubble effect discovered by Edwin Hubble in 1929, but tracing the temporal sequence of events is difficult.

156 Quoted from Peter Day, *The Philosopher's Tree, A Selection of Michael Faraday's Writings*, p. 101 ff.

Physical discoveries will thus not normally be blissful events for the axiomatist, least of all if made by others. On the contrary, they tend to be troublesome. Inevitably they mean work, compelling him at least to adapt to new findings his old theories. At worst they can even wreck his life's ambition. He has cause to stand up in defence against them – with all the means at his disposal!

III

That must have been Einstein's problem too. Remarkable events in his life show that he only reluctantly took notice of discoveries that did not fall in with his theories. The truly great discoveries carried away even most theoretical physicists in the end. Einstein however was palpably averse to recognizing them if they did not agree with his own thoughts. One of them he ignored persistently.

(1) When Einstein first published his general theory of relativity in 1916, observations had already shown that the spectrum of sunlight differs from that of light reaching us from more distant stars[157]. The phenomenon is called 'redshift' now, but little was known of it at that time because accurate measurements had not yet been made. In his fourth Princeton lecture on the general theory of relativity (1921) Einstein explained the redshift like this[158]:

> 'The rate of a clock is accordingly slower *the greater is the mass of the ponderable matter in its neighbourhood.* We therefore conclude

---

157 Einstein mentioned that in a footnote on p. 820 to *Die Grundlage der Allgemeinen Relativitätstheorie* (The Foundation of the General Theory of Relativity) Annalen der Physik vol. 49 (1916), p. 769–822.

158 Albert Einstein, *Die Grundlage der Allgemeinen Relativitätstheorie*, Annalen der Physik vol. 49 (1916), p. 769, 775; *The Meaning of Relativity* (1921), p. 97; for details see also my *Popper versus Einstein*, p. 91–95.

that spectral lines which are produced on the sun's surface will be displaced towards the red, compared to the corresponding lines produced on the earth, by about $2 \cdot 10^{-6}$ of their wave lengths.' (My italics).[159]

Going by that explanation the redshift would have been correlated to *mass*. The greater the mass of a distant star, the stronger should have been the redshift in light coming from it.

In 1929, however, the American astronomer Edwin Hubble not only observed that the shift always was in the direction of lower frequencies but also that it showed a continuous increase correlated to *distance*. Einstein's interpretation became untenable thereby because aligning it with Hubble's observations would have meant that the size of stars must increase in proportion to their distance from the earth. And that would have clashed with observations showing that stars with equal redshift can have different sizes.

Hubble himself had taken a different approach. He had interpreted the redshift by the Doppler principle as indicating a flight movement of remote stars, hence of an expanding universe[160]. Einstein accepted that only reluctantly because it not only meant that his own explanation had been wrong but also that his general theory of relativity had been mistaken and that the mistake was not in details only. It was crucial.

He made a correction to the general theory but it must have taken him long to get it written down. He never published it in his lifetime as far as I can see. It was published posthumously in 1956 as an appendix to the first expanded German version of his Princeton lectures, and it seems to have remained unnoticed in the English-speaking world of theoretical physics. I could find no English publication of Einstein's Princeton lectures including that appendix. It seems not to exist[161].

---

159 *The Meaning of Relativity*, p. 97.

160 Edwin Hubble, *The Realm of the Nebulae* (1935). - I do not believe that interpretation to be correct. Mine is that light loses energy on its way through space (*Popper versus Einstein*, p. 183). That explains also the phenomenon of entropy. It is a consequence of returning to ether theory, which I believe to be inevitable as we will see in *Chapter 11*.

161 The English title of the Princeton lectures is *The Meaning of Relativity*. The first two

In that appendix Einstein adapted his general theory to Hubble's discovery by introducing a new constant h[162]. It now is named 'Hubble constant' although measurements known in Einstein's time and today hardly justify assuming the correlation between redshift and distance to be unvarying[163]. Without a constant however, the general theory would have broken down altogether because no calculation would have been possible. The general theory of relativity would not have applied to the universe any more. By introducing the abstract constant h, standing for the Hubble expansion, Einstein tried to rescue his theory but presupposed a knowledge not existing. Except for dishonesty, of which I see no indications on his side, the only plausible explanation is his unswerving belief that the universe *must* be calculable. In *Chapter 8* we will see that Max Planck held similar views[164].

German editions had been published under the title *Vier Vorlesungen über Relativitätstheorie*. The appendix was added in the 3rd German edition (1956). Its title is *Zum kosmologischen Problem* (On the cosmological Problem). It is not identical with Einstein's short paper *Zum kosmologischen Problem der allgemeinen Relativitätstheorie* (On the cosmological problem of the general theory of relativity), published in 1931 (Sitzungsberichte der Königlich Preußischen Akademie der Wissenschaften [1931], p. 235–237). For the latter see Harry Nussbaumer's report *Einstein's conversion from a static to an expanding universe*, European Physics Journal – History, 39 (2014) p. 37–62. In that 3rd German edition of the Princeton lectures (1956), the German title was changed to *Grundzüge der Relativitätstheorie*. The effect of changing the title at the same time as introducing a fundamental change of the General Theory of Relativity appears to have been that this fundamental change remained unnoticed in the English-speaking world of theoretical physics. At any rate, I could find no English translation of that appendix.

162 *Grundzüge der Relativitätstheorie*, Appendix I to 6th ed. 1956, p. 107 ff., 116–119, 125–130.

163 Assuming the correlation between redshift and distance to be unvarying presupposes that the speed of light in the atmosphere of the earth is the same as in outer space. It would conflict with Einstein's premise that (only) the speed of light *in a vacuum* is constant. Besides, there is no empirical evidence supporting that assumption. In *Chapters 7, 11* and *12*, we will see that it conflicts also with other empirical findings.

164 See Popper's report in *Unended Quest*, p. 127f. - The development described in the text is one of the reasons why I think that if there was dishonesty in the history of the theory of relativity, then it was not Einstein's but that of his followers or of his publishers.

(2) My second example concerns the special theory of relativity. Einstein had based it on the hypothesis that the speed of light in a vacuum is constant, independent of the motion of its source. I will discuss it in more detail in *Chapter 9*. In his first presentation of the special theory Einstein had introduced that hypothesis as a premise without giving further explanations[165]. Beginning like that was characteristic of his heuristic approach as we saw. He would introduce some premise without explaining it, and would trust that applying it would demonstrate its worth. After that beginning, anyone believing in experiments could still think that putting to the test in targeted experiments Einstein's hypothesis was what mattered now.

That must have been where the French physicist Georges de Sagnac (1869–1926) believed matters to stand[166]. He was one of the rare *experimental critics* of the special theory of relativity in its early years. From 1910 to 1913 he carried out a series of interesting experiments.

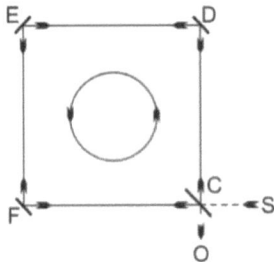

Figure 1, Sagnac's experiment, schematic diagram: The entire setup is turning on its vertical axis. The light beam starting in *S* is semi-reflected to *D* and *F* in *C*, and then reflected again in *D*, *E* and *F*. Two split beams pass through *C, D, E, F* in opposite directions, and then are partly reunited at *C* and reflected to *O*.[167]

165 Einstein, *On the Electrodynamics of Moving Bodies*, Introduction. More details in *Chapter 7*, below.

166 Sagnac, *L'éther lumineux, démontré par l'effet du vent relatif d'éther*, Comptes rendus, vol. 157 (1913), p. 708 ; *Sur la preuve de la réalité de l'éther lumineux*, Comptes rendus vol. 157 (1913), p. 1410 ; J. de Phys. (1914). Pt. 4, p. 177.

167 The diagram is quoted from Al Kelly, *A New Theory on the Behaviour of Light'*.

Sagnac fastened a source of light (*S*) on a rotating disc and let it rotate with the disc. A beam emitted by that source was split in a semi-reflecting mirror and then reflected along the periphery of the disc (see *Figure 1*). Part of the beam thus propagated in the direction of rotation (C,D,E,F) and the other part against it (F,E,D,C). The two split beams then reunited in the semi-reflecting mirror at C and were tested for interferences. Both the source of light and the measuring device itself (O) were on the rotating disk. The experimental set-up was rigid in itself but pivoting on its vertical axis.

Einstein's special theory assumed the vacuum speed of light to be always constant, independent of the state of motion of its source. It did not say why carrying out experiments in normal atmosphere should make a difference. Sagnac carried out his experiments in normal atmosphere. Nevertheless, their environment being the same, the light beam spreading in the direction of rotation and that spreading against it should at least in principle both have had the same velocity. If the special theory had been right, then there should have been no significant interferences after reuniting those beams. In the entire set-up there was no motion of the source of light relative to the orbit of the light ray, nor relative to the measuring instrument. Apart from the velocity of light itself the only motion was that of the rotating experimental set-up, including the light source, *relative to the environment* of the set-up.

In fact however, notable interferences showed even at quite low rotational velocities, and they increased and decreased in proportion to them. They indicated that the speed of light propagating in the direction of rotation must be different from that of light propagating against it[168]. Nobody

---

168 My German text (*Albert Einstein oder: Der Irrtum eines Jahrhunderts*, p. 89) here has the sentence 'Das mit der rotierenden Lichtquelle beschleunigte Licht schien von der umgebenden Atmosphäre wieder gebremst zu werden.' (In English: 'The light, which had been accelerated by the rotating light source, appeared to have been decelerated again by the surrounding atmosphere'). I have omitted that sentence here because it was wrong. It was not in my oldest version of the German text (2003), and I cannot have been at my best when I put it in later. For, clearly, the cause of the interferences observed by Sagnac did not lie in a difference of the speed of light relative to the surrounding atmosphere, but in the differences between the speeds of light emitted in the direction of rotation and that emitted in the opposite direction.

proposed a better interpretation to this day, nor did anybody explain in any other way the correlation between rotational velocities and interferences. It seemed obvious that Einstein's hypothesis of the speed of light in a vacuum being constant, independent of the motion of its source, had not stood up to experimental test. In view of that, and by normal standards of experimental physics, theoretical physics should either have accepted that result or embarked on a discussion of special relativity.

Nothing of that kind happened however, although German critics of the theory of relativity did not overlook Sagnac's experiment even in his time and tried to draw attention to it[169]. Einstein himself simply ignored it and other theoretical physicists did likewise. That situation persists to this day. Although any physicist can repeat the experiment at any time, modern textbooks for students of special relativity still treat it as non-existent. It would probably have been forgotten altogether if the late Irish physicist Al Kelly had not snatched it from oblivion in 1995, thereby prompting the French theoretical physicist J. P. Vigier almost immediately to propose an amendment to the special theory leading to new contradictions elsewhere[170]. No theoretical physicist seems to realise that the problem is not in adjusting in details the special theory of relativity, but that its all-important basic premise of the speed of light being constant, independent of the motion of its source, has been refuted empirically.

---

169   One of them was H. Fricke in *Der Fehler in Einsteins Relativitätstheorie* (1920).

170   Kelly, *Challenging Modern Physics* (2005), p. 113f.; the same, *A New Theory on the Behaviour of Light*, The Institution of the Engineers of Ireland, Monograph No. 2; Vigier, *New non-zero interpretation of the Sagnac effect as a direct experimental justification of the Langevin paradox*, Phys. Lett. A (1997), p. 75 ff. - Vigier assumed a 'nonzero rest mass' of the photon there. According to the special theory of relativity, its mass at the velocity c would be infinite then. I commented on that in *Popper versus Einstein*, p. 59, 66 f. Going by Kelly's report (*Challenging Modern Physics*, p. 114), Vigier himself expected that 'not many would swallow' his 'amendment' to the general theory.

Another problem of epistemology is raised by the necessity of *interpreting* observations and experiments. For the progress of science it is at least as important as the discoveries from which progress can only result. It also deserves careful consideration therefore.

(*1*) Observation always is the result of choice, as we saw in *Chapter 2, II.* No observation is important of its own. It can only be important-for-something and must start from an interpretation in the light of some theory. Similarly, experiments and discoveries can also only be important-for-something. Recognizing their importance requires an interpreting theory. I begin by stating examples.

(*a*) In 1789 the Italian physician Luigi Galvani (1737–1798), in search for an explanation of biological life, was dissecting frogs when he made an observation which others probably had made long before him. That the extremities of some creatures will continue twitching after being separated from their bodies must have been known from time immemorial but never been considered important.

Galvani's interpretation of that observation was new however. He surmised that twitching muscles might have to do with electricity somehow, and that made him bring the frog legs near a device for generating static electricity and observe the effect of leaping sparks on their twitching[171]. Only that conjecture, later to be modified and specified by Alessandro Volta (1745–1827), made his observation the most important discovery after the invention of the steam engine. It revealed that electricity is not just a property of some specific materials, such as the resin used in electrifying machines, but in fact *omnipresent*. It also demonstrated that believing in the existence of *invisible* forces of nature need not be superstition

---

171 See Teichmann, *Das Werk von Galvani und Volta*, in v.Meÿenn, *Die Großen Physiker*, vol. I, p. 266 ff.

but may withstand even the most rigorous rational criticism if exposed to experimental tests. The stupendous development of science unleashed by Galvani's and Volta's discoveries did not result from the observations they had made but from the hypothesis they framed for interpreting them. It explained those observations as effects of electricity, and framing that hypothesis had been the result of their personal power of imagination. It was the theory of electricity, not the observation of twitching frog legs, that led to all those new applications and discoveries in the field of electricity which have not come to an end yet.

(b) Nuclear fission is another striking example from more recent times. General opinion assumes Otto Hahn and Fritz Strassmann to have discovered it in 1938. Some also mention Hahn's former assistant Lise Meitner who had left Germany earlier that year due to the persecution of Jews there.

Telling a different story would be possible however. In 1934 Enrico Fermi (1901–1954) at Rome had started bombarding the nuclei of uranium with neutrons and other particles in order to convert them into heavier elements. In a review of his work the German chemist Ida Noddack ventured the hypothesis that bombarding with neutrons might split heavy nuclei into large fragments[172]. The professional world did not take her seriously however, because her conjecture stood in open conflict with the theory almost uncontested at that time. Most physicists still believed the nucleus to be indivisible then[173]. In October 1937 Niels Bohr, Nobel laureate for Physics of 1922, and Fritz Kalckar published a voluminous paper dealing with that question[174]. They argued that for quantum-theoretical reasons, bombarding the nuclei of uranium with neutrons, protons or al-

---

172   Ida Noddack, *Über das Element 93*, Angewandte Chemie, vol. 47 (1934), p. 653 f.

173   Karl-Erik Ziemen mentions that in his introduction to Otto Hahn, *Erlebnisse und Erkenntnisse* (1975), p.11; see also Walther Gerlach, *Otto Hahn Ein Forscherleben unserer Zeit*, ed. by Dietrich Hahn (1984), p. 77. - Not many seem to have interpreted Rutherford's discovery that alpha rays would transform nitrogen into oxygen and hydrogen as a case of nuclear fission at that time (1919), but Einstein apparently saw the implication and realised its importance (see Moszkowski, *Einstein* [1921], p. 48).

174   Niels Bohr/Fritz Kalckar, *On the Transmutation of Atomic Nuclei by Impact of Material Particles*, § 7 (p. 31).

pha particles could only emit in 'an evaporation like process of the nuclear matter' light nuclear fragments, hence again only neutrons, protons or alpha particles. That same year Irène Joliot-Curie and Paul Savitch at Paris also bombarded the nuclei of uranium with neutrons and found among the remains a substance with a half-life period of 3.5 hours but could not identify it. All these facts are undisputed historically[175].

Someone might argue therefore that nuclear fission had in fact been discovered in 1937. In 1934 Ida Noddack had framed the hypothesis that neutron bombardment could split heavy nuclei into large fragments. And the experiments made by Irène Joliot-Curie and Paul Savitch in 1937 had confirmed her surmise. Acknowledging physical discoveries usually requires no more than that. Hahn and Strassmann would then not have discovered nuclear fission in 1938 but only provided a more accurate confirmation of it by identifying the fission product as an isotope of barium .

Nevertheless it is correct to consider the two chemists Otto Hahn and Fritz Strassmann the discoverers because the discovery did not consist in doing it but in finding out what had happened. Anyone reading the letters which Otto Hahn and Lise Meitner exchanged between November 26, 1938 and April 22, 1939 will realise that[176]. Others before them had already bombarded uranium nuclei with neutrons. The difficulty Hahn and Strassmann had to overcome consisted in *identifying* the remainders obtained in that process. They were the first to approach it systematically

---

175 Helmut Rechenberg, *Lise Meitner und Otto Hahn, Irène Joliot-Curie und Frédéric Joliot* in v.Meÿenn, *Die Großen Physiker*, vol. II, p. 210 ff, 219 ff.

176 Otto Hahn, *Erlebnisse und Erkenntnisse* (1975), p. 75–129. – Apart from those letters, I am relying for the facts related in the text also on Noddack's article of 1934, on reports by Rechenberg and Kragh in *Große Physiker* (pp. 219, 362), and on a personal communication by Otto Hahn's grandson Dietrich Hahn in 1997. Noddack never claimed priority for the *discovery* of nuclear fission but she complained at not having been mentioned in that context with her surmise of its *possibility* (see Otto Hahn's letter to Lise Meitner of March 29, 1939, in Otto Hahn, *Erlebnisse und Erkenntnisse* [1975], p. 115–116). Popper seems to have believed that Lise Meitner and her nephew Otto Robert Frisch were the first to develop the idea of interpreting as a case of nuclear fission Hahn's experiment. He describes that interpretation as having been new then (*Alles Leben ist Problemlösen* p. 260). In view of the evidence just mentioned I think he must have been mistaken.

by making long sequences of experiments and writing down and comparing all their results[177]. Thus they were the first to open their minds to facts instead of clinging to the dogmas of atomism and quantum theory. They also were the first to seriously oppose the Bohr-Kalckar theory unchallenged until then. Reading those letters shows that their self-doubts persisted even after they had published their experiments, and that overcoming those self-doubts was almost their greatest problem. If they were to trust in results showing that bombarding the uranium nucleus with neutrons will create isotopes of barium, then they had to assume the nucleus to have split down the middle as it were. They published that conjecture for scientific discussion and investigated and specified the experimental conditions under which other scientists could repeat their observation and in fact repeated it worldwide. By stating that counter-theory and by encouraging others to probe if by experiment they were the first to take the risk of being refuted themselves. Framing the counter-theory and specifying the conditions required for testing it was the discovery, not making observations without understanding them.

(c) The example throws some light on the principle underlying scientific discoveries. When Copernicus put forward his heliocentric theory about in 1514 the visible movements of the planets had been known for centuries, most of them for millennia. Those very movements had already inspired Ptolemy in his attempt at explaining the cosmos by spheres revolving around the earth. The observations were no breakthrough therefore. Even after Copernicus, when the invention of the telescope in Holland had further improved the accuracy of observations, the Danish astronomer Tycho Brahe (1546–1601) still tried to align Ptolemaic theory with the visible motions of the planets by assuming eccentrical and epicyclical movements. Not observation turned the universe upside down but the heliocentric theory which moved the sun to the centre of the universe and

177 Walther Gerlach gave a comprehensive summary including photostats in *Otto Hahn, Ein Forscherleben unserer Zeit*, ed. by Dietrich Hahn (1984), p. 78–100. Heisenberg's report in *Erinnerungen an die Entwicklung der Atomphysik in den letzten 50 Jahren* (in *Deutsche und jüdische Physik* p. 187 ff, 204) also confirms that view.

put the earth on one of its orbits, degrading her thus to a place among the planets. That theory enabled Kepler to find the laws of the ellipse (1619) which entailed many other discoveries, among them those of Neptune (1846) and Pluto (1930). Copernicus was the first to interpret the visible motions of the planets in the light of his heliocentric theory. That was his revolutionary achievement.

There are many examples of discoveries but those just mentioned ought to have shown the aspect relevant for the theory of science. Every experiment must be interpreted, and scientific observation can be recognized *as a discovery* only in the light of that interpretation. The discovery does not consist in making observations but in proposing the interpreting theory and testing it if possible. Only theory reaches beyond the well-known experiment and thus yields new information. The progress of knowledge brought about by discoveries does not originate in observation either, but in the fact that the interpreting theory will provide information reaching beyond observation.

(2) That has far-reaching consequences for the efforts of our axiomatist to whom we must now return. It means that whether a discovery agrees with his thoughts or will endanger his goal of research depends not only on observations but also their interpretation. That leaves him room for hope. Even unexpected discoveries need not threaten his life's work if he can interpret them as agreeing with his own theory. The alternative he faces is obvious. He can either change his theory by dropping old axioms or recognizing new ones. His goal of making 'the irreducible basic elements of science as simple and few as possible' will then be further removed than it had been before the discovery. The work of his life will be in jeopardy. Or he can try to interpret the new discovery as corresponding with his own theory. If that succeeds, then the discovery will cause him no problems but might even gain him considerable profit. Given sufficient power of persuasion he can now claim that it had brilliantly confirmed his prediction. He can almost feel having been the discoverer himself.

Clearly, the latter way of dealing with new discoveries is by far to be preferred. That must have been why it became the standard attitude of theoretical physics in defence of relativity as we will see further down. At this point nuclear fission must once more serve as an example for illustrating the argument and preserving the order of the essay.

(a) Apart from Ida Noddack whose proposition few if any took seriously[178], no scientist but Einstein seems to have believed in the possibility of nuclear fission[179]. All those engaged in the neutron experiments, Enrico Fermi and Irène Joliot-Curie at any rate, but also Lise Meitner, Otto Hahn and Fritz Strassmann, were in fact starting from different assumptions and pursuing different goals[180].

As mentioned, Fermi had discovered in 1934 that bombarding with slow particles could convert the nuclei of certain atoms. His aim was *adding* more particles to the nucleus of uranium, the heaviest then known element. He hoped to produce so-called transuranic elements of even higher mass number and thus to prolong the list of hitherto unknown elements. Lise Meitner, working in Otto Hahn's team, heard of those experiments and persuaded Hahn to tackle the same problem[181]. Irène Joliot-Curie at Paris also made experiments to that effect.

In all those experiments nuclear fission neither was intended nor expected, rather on the contrary. They did not aim at splitting the uranium nucleus in smaller fragments but at increasing it in size by adding to it more particles in order to produce transuranic elements. Creating materials with lower mass numbers by shooting off small components of the nu-

---

178 See *sec. 1b*, above.

179 Fölsing (p. 795) quotes from a newspaper article dated July 25, 1920 where Einstein deemed possible that Rutherford's discovery of nuclear transmutation (1917) might open new sources for gaining energy. Going by Alexander Moszkowski's report (*Einstein*, 1921, p. 48), Einstein may have realised also that it had been a case of nuclear fission.

180 Heisenberg, *Erinnerungen an die Entwicklung der Atomphysik in den letzten 50 Jahren* (in *Deutsche und jüdische Physik* p. 187 ff, 204), Rechenberg *Lise Meitner und Otto Hahn, Irène Joliot-Curie und Frédéric Joliot* in v.Meÿenn, *Die Großen Physiker*, vol. II, p. 210 ff, 219 ff.

181 Rechenberg, *Lise Meitner und Otto Hahn, Irène Joliot-Curie und Frédéric Joliot* in *Die Großen Physiker*, ed. by Karl v. Maÿenn (1997), vol. II p. 210, 219.

cleus also seemed conceivable. Real nuclear fission however, splitting the nucleus near its middle as if cleaving it with an axe, seemed impossible. As mentioned above, Niels Bohr and Fritz Kalckar had 'proved' that in October 1937 on quantum-theoretical grounds[182]. One of the most famous theoretical physicists of that time had thus excluded the possibility that an atomic nucleus might disintegrate into two or more nuclei with much smaller mass numbers.

Fission, however, was what Otto Hahn and Fritz Strassmann observed when they bombarded with neutrons uranium with mass number 235 and consistently found barium with mass number 138 among the remains. As in the case of the Sagnac effect, someone among the leading theoretical physicists should actually have realised then that they had failed, and that their approach to the problem had been misguided.

It was a humiliation of the first order for theoretical physics that two chemists should have made a discovery of such unparalleled importance in the field of physics shortly after one of the most prominent Nobel-laureates of theoretical physics had declared it to be impossible. Throughout the 20th Century no other discovery changed the worldview of science and of the public as fundamentally and in so short a time as nuclear fission and its consequences did. At a single stroke and beyond all argument it made apparent by hard experimental facts that matter is not passive, as classical physics had assumed and as quantum theory and the theory of relativity had presupposed, but is a dynamic *process* even in its smallest components. It should also have changed the worldview of physics because no physical theory had even remotely anticipated anything like that. Theoretical physics was not just off after Lise Meitner's forced emigration. It was stumped.

(*b*) It seems that in that embarrassing situation Einstein himself came to the rescue by pointing to $e = mc^2$, a formula he had also published in 1905 shortly after his special theory of relativity[183].

---

182   Bohr/Kalckar, *On the Transmutation of Atomic Nuclei by Impact of Material Particles*, § 7 (p. 28–34, particularly p. 31).

183   At the time of writing *Albert Einstein oder Der Irrtum eines Jahrhunderts*, my infor-

According to Moszkowski's report, Einstein was the only theoretical physicist having no reason to be surprised by Hahn's discovery because his own thoughts had not been with the mainstream[184]. In 1919, when Rutherford had discovered that bombarding with alpha rays will transform nitrogen into oxygen and hydrogen, Einstein had already interpreted that as a case of nuclear fission and had brought it in connection with his formula $e = mc^2$. The formula had remained almost unnoticed until then, but after the discovery of nuclear fission (1938) and especially after the atom bomb many came to consider it the most important formula of the theory of relativity. The equivalence of mass and energy which Einstein had postulated there was now to explain that smallest quantities of matter could release huge energies. Due to that interpretation theoretical physicists could claim the unexpected outcome of Hahn's experiment to stand in beautiful harmony with the theory of relativity. $E = mc^2$ remains Einstein's most famous formula to this day.

I used to doubt this interpretation but now believe it to be correct historically. Moszkowski published his Einstein-book in 1921. Later events could not have influenced his report therefore. I had not read his book when I published *Albert Einstein oder Der Irrtum eines Jahrhunderts* (2009), and reading it later did not influence my assessment of $e = mc^2$ as I will explain in *Chapter 9, VI*. In my view of Einstein's attitude in developing the formula however, it caused a great change because only Moszkowski's report made me realise that the atom bomb, which Einstein could not have known at the time, had too strongly influenced my own interpretation of those events. But it still surprises me that the connection between $e = mc^2$ and nuclear fission could continue to be believed after the bomb (1945).

---

mation, given me orally by Dietrich Hahn in 1997 as a recollection of a communication made to him by his grandfather Otto Hahn (1879–1968), was that Bohr had been the first to mention $e = mc^2$ in the context of nuclear fission. Going by Moszkowski's report, however (*Einstein*, 1921, p. 48), which I did not know then, it seems more likely that it was Einstein himself. At present, at any rate,, most will consider the connection between $e = mc^2$ and nuclear fission to be self-evident (see, for example French, *Special* Relativity, M.I.T. textbook, p. 18). I will discuss that in *Chapter 9, VI*.

184 Alexander Moszkowski, *Einstein* (1921), p. 48.

We will see in *Chapter 8* that Einstein's derivation of $e = mc^2$ was not only circular but fallacious also for other reasons. After the bomb, anyone with a critical mind and forming opinions of his own instead of blindly adopting those of others will easily see that $e = mc^2$ cannot possibly serve as an explanation of nuclear fission. It cannot even serve as a *description*; its very shape already shows that. If c, standing for the speed of light, is constant as the theory of relativity claims, then c^2 is constant too. Other than some might assume at first glance, $e = mc^2$ is no formula of a parabola therefore, describing by an exponential equation an increase of energy as in an explosion. All it describes is a linear correlation of mass and energy. In a system of coordinates it would be depicted as a straight line with its slope angle depending only on the scales selected. The atom bomb however revealed something quite different. It showed that matter could be dynamite, *creating* huge energies with stupendous velocity. That was in some respects just the opposite of the harmless linear equivalence of mass and energy which Einstein had postulated in $e = mc^2$. It also was why no theoretical physicist had even remotely anticipated that effect[185].

For experimental physics all that was to no avail. After $e = mc^2$ had been linked with nuclear fission nobody was interested anymore in the derivation or the interpretation of the formula. It had ostensively saved the honour of theoretical physics, and many believe still today that there is an immediate relationship between the discovery of nuclear energy and Einstein's theory of relativity, a myth studiously fostered by some of his adulators[186].

---

185 Not even Einstein had anticipated it. In his paper *Über einen die Erzeugung und Umwandlung des Lichts betreffenden heuristischen Gesichtspunkt* (On a Heuristic Point of View Concerning the Production and Transformation of Light), Annalen der Physik vol. 17 (1905), p. 132 ff., he assumed energy to be inherent in any kind of matter, which would mean that one kilogramme of water contains as much energy as one kilogramme of coal. I will discuss that in detail in *Chapter 9, VI.*

186 Fölsing's representation (*Einstein – Eine Biographie*, p. 794, 795) is symptomatic of the attitude described. In the context of the beginning of the war (1939) he claims that Einstein had immediately understood the military, political and physical implications of nuclear fission. For substantiating that he refers in the text and in a note to much earlier papers relating not to Einstein's paper of 1905 but to Rutherford's discovery of the possibility of transforming the nucleus (1919).

(3) When discussing experiments concerning the theory of relativity it sometimes is not easy to find out whether the results of experiments had been adapted to the theory or the theory had been adapted to those results. In the experiments on time shifting through velocity carried out by J. C. Hafele and R. E. Keating in 1971 both seems to have happened.[187]

(a) It is a well-known allegation of special relativity that time will run more slowly in a moving system than in a system at rest. The effect is now called 'time dilation', and adherents of the special theory believe it to follow by cogent inference from valid premises. Assuming that to be true, it would raise many questions. One of them is whether a space traveller returning to earth after a long voyage through space would have remained younger than his twin brother who had stayed here, or whether the deceleration at the return of the vessel would compensate the alleged time effect on the traveller. The formulae of special relativity do not answer that question because they apply only to constant motion on straight lines. If time dilation did exist, then calculating the time shift for accelerated or decelerated motion would have to rely on the general theory which hardly permits to infer from it any specific results. Besides, most physicists not even pretend being able to understand it.

Due to those difficulties it was highly controversial for a long time what effects on time dilation would follow from the theory of relativity in cases of accelerated or decelerated motion. An experiment carried out in 1971 by Hafele and Keating was to bring the clarity desired. They let fast commercial aeroplanes carry four caesium clocks around the earth near the equator, two each in eastern and in western direction, and then compared them with stationary clocks that had been synchronized with

---

187   The discussion, following in the main text, of the experiments on time dilation made by J.C. Hafele/R.E.Keating differs from my German text in *Albert Einstein oder: Der Irrtum eines Jahrhunderts*. My interpretation of the *results* of those experiments remains unchanged. I must correct my interpretation of the *events* in one point however, because at the time of writing the German book I had not read Hafele's paper *Relativistic Time for Terrestrial Circumnavigations* (AJP, vol 40[1972], p.81–85) where he had predicted almost precisely the time differences he later observed.

them before the flights[188]. Both transports took more than 24 hours *in flight*. That will be important for interpreting the experiment. The eastward transport took 41.2 hours, the westward transport took 48.6 hours.

It turned out there was a significant effect on the indications of clocks which had travelled compared to those remaining stationary. Insofar the result seemed to agree with special relativity. However, the clocks did not show the deceleration in both directions predicted by Einstein but only in eastward direction, whereas the clocks travelling westwards against the direction of the earth's rotation had been accelerated. Under the rule of relativity that should never have been permitted to happen because the special theory contends that the time of the moving system will *always* be slower than that of the system at rest, irrespective of the direction of relative motion. In his first presentation of the theory Einstein had stated that very clearly in the following words:

'If one of two synchronous clocks at A is moved in a closed curve with constant velocity util it treturns to A, the journey lasting t seconds, then by the clock which has remained at rest the travelled clock at its arrival at A will be ½ tv^2/c^2 second *slow*. Thence we conclude that a balance clock at the equator must go *more slowly*, by a very small amount, than a precisely similar clock situated at one of the poles under otherwise identical conditions.'[189] (my italics).

This shows that under no circumstances whatever could an acceleration of time be reconciled with Einstein's own interpretation of his special theory.

---

188 J. C. Hafele/R. E. Keating, *Around-the-World Atomic Clocks Predicted Relativistic Time Gains*, p. 166; the same in *Around-the-World Atomic Clocks Observed Relativistic Time Gains*, p. 168. - A. P. French (*Special Relativity* [ed. 1997] p. 97 ff.) reported of another experiment apparently made in 1963. The type of that experiment was different but one of the logical mistakes underlying French's interpretation of it was the same as that made by Hafele/Keating. I will mention it in footnotes as we go along.

189 Einstein, *On the Electrodynamics of Moving Bodies*, end of § 4.

(*b*) Nevertheless, Hafele saw things differently. In a paper published *before* carrying out those experiments with airborne clocks he predicted that the clocks travelling eastwards would be decelerated whereas those travelling westwards would be accelerated. His explanation for that prediction was that that the offset between the clocks was 'a unique (invariant) quantity that is independent of the frame of reference of an observer who measures it'[190].

The explanations given by Hafele and Keating *after* the experiments were less enigmatic. They wrote[191]:

'Consider a view of the (rotating) earth as it would be perceived by *an inertial observer looking down on the North Pole from a great distance*. A clock that is stationary on the surface at the equator has a speed $R\Omega$ relative to nonrotating space, and hence runs slow relative to *hypothetical coordinate clocks of this space* in the ratio $1-R^2\Omega^2/2c^2$, where R is the earth's radius and Ω its angular speed. On the other hand, a flying clock circumnavigating the equator near the surface in the equatorial plane with a ground speed v has a coordinate speed $R\Omega + v$, and hence runs slow with a corresponding time rate $1- (R\Omega + v)^2/2c^2$. Therefore, if $\boxtimes$ and $\boxtimes_0$ are the respective times recorded by the flying and *ground reference clocks* during a complete circumnavigation, their time difference, to a first approximation, is given by $\boxtimes-\boxtimes_0 = (2R\Omega v + v^2)\,\boxtimes_0/2c^2$. Consequently, a circumnavigation in the direction of the earth's rotation (westward, $v>0$) should produce a time loss, while one against the earth's rotation (eastward, $v<0$) should produce a time gain for the flying clock if $v \sim R\Omega$.' (My italics).

(*c*) The logical mistake in that reasoning is obvious. When deriving their formula for calculating time dilation, Hafele/Keating had assumed the perspective of 'an inertial observer looking down on the North Pole from a great distance'. My italics show that. The observer himself was to be at

190 Hafele, *Relativistic Time for Terrestrial Circumnavigations* (AJP, vol. 40[1972], p.81, 83.
191 Hafele, *Around-the-World Atomic Clocks: Predicted Relativistic Time Gains*, Science vol.177 (1972), p. 166.

rest therefore because otherwise he would not have been 'inertial'. In their experiment however, they did not comapare the airborne clocks with the inertial observer but with 'ground reference clocks' resting near the equator and therefore *rotating with the earth*. In both directions the transport took more than 24 hours in flight for one circumnavigation. The rotational speed of the earth's surface was higher at all times than the speed of the planes over ground. Even the planes *flying* in eastwards were therefore *rotating* westwards. Relative to the 'inertial observer looking down on the North Pole from a great distance' their speed could have been negative ($v < 0$) only if the observer himself were spinning on the prolongation of the pole axis, but in that case he would not have been 'inertial'. The calculations made by Hafele/Keating were fallacious because they did not take into account the speed of the stationary clocks caused by the earth's rotation.

That is not the only mistake they made. The more fundamental question is what their experiment had to do with the theory of relativity. Clocks *are* not time but instruments for measuring it. A good clock should indicate what we *make* it for, but any clock can be fast or slow. If we make it for comparing the position of the sun to that of the earth, then a clock indicating differently from that is wrong; and if we make it for comparing the position of the moon to that of the earth, then a clock indicating differently from that is wrong. If a clock correctly compares the position of the sun to that of the earth as long as it is stationary, but runs differently after orbiting the earth, then something must have happened to the clock on that orbit. If the effect can be repeated, then the only possible inference is that the clock changed its rhythm under the influence of the voyage[192].

---

192 The same confusion of time and the instrument used for measuring it shows also in A. P. French's discussion of time dilation (A. P. French, *Special Relativity*, ed. 1997, p. 101–104). French there reports of an experiment using mesons as an instrument for measuring time and yielding different results at the top of Mount Washington than at sea level. Before attributing that to a difference of time measured by comparing the position of the earth to that of the sun he would have had to make sure therefore that their behaviour on the top of Mount Washington would be *the same* as that at sea level.

That is as obvious as can be. The next question must be why it changed[193]. I will discuss that in *Chapter 9*.

(*d*) The most miraculous aspect of the Hafele/Keating story however is this. Before carrying out those experiments Hafele *predicted* a deceleration of the clocks travelling eastwards and an acceleration of the clocks travelling westwards, and he stated a formula for calculating the offset to be expected. He based that on an interpretation of the theory of relativity which was directly opposed to Einstein's own interpretation as the text quoted from Einstein's paper shows (above, *a*). He also based it on a logical mistake as we saw (above *c*). Yet the outcome of the experiments confirmed his predictions with remarkable precision.

The late Irish physicist Al Kelly made careful investigations for finding the explanation of that miracle, comparing the protocols of the flights with the data published by Hafele/Keating in their paper[194]. According to his report the results actually measured were not the same as those published. Yet they also showed a deceleration of the clocks travelling eastwards and an acceleration of the clocks travelling westwards. Kelly's own interpretation was that the data published in the Hafele/Keating paper after the flights had been adapted to the results predicted by Hafele in his paper published before the flights. And his final inference was that Hafele, who was working for the U.S. Naval Observatory at the time, must have known before carrying out his experiment that there would be an acceleration in the clocks travelling westwards, but needed the adaption for con-

---

193 In the experiment made by Hafele/Keating the most obvious explanation seems to me that ether wind had caused the effect. In *Chapter 11*, we will see that moving against the direction of the earth's rotation will increase the velocity of matter relative to ether. - In French's experiment (see previous note) the difference of altitude may have caused the mesons to behave differently. Under the assumptions made in *Chapter 11* that too is to be expected.

194 For the experimental results see Kelly's report doubting the correctness of protocols in Kelly, Al: *Challenging Modern Physics* (2005), p. 31–34; 265–278 *Reliability of Relativistic Effect Tests on Airborne Clocks* (The Institution of the Engineers of Ireland, Monograph No. 3). For logical mistakes in the way they were interpreted see my ‚Popper *versus* Einstein', p. 69 ff.

firming the formula he had proposed in his previous paper. Kelly's final conclusion was that 'the whole story describes a shameful episode'[195]. He did not mention that Hafele/Keating had also adapted the interpretation of the theory of relativity itself to their empirical data.

Following Newton's great example I leave 'to the consideration of my readers' what to make of all this. The most bewildering aspect of the Hafele/Keating story is that modern textbooks still present it to unsuspecting students as a brilliant confirmation of the theory of relativity, proving that speed will in fact influence the flow of time[196].

V

Those examples raise the question of what point there might be in making experiments as long as theoretical physics remains in the state just demonstrated. How can we learn anything from experiments if we acknowledge only results confirming our preconceived views?[197] How can we test

195   Kelly, *Challenging Modern Physics* (2005), p. 272–275. – In view of what Kelly wrote about me there (p. 260) I would like to mention that he had very kindly helped me with my *Popper versus Einstein* (1998) by giving me some of the interesting information quoted there, and by looking after my English. I think he was not quite convinced of the approach I had taken in my book, and his own last book shows the reason for that. My approach comes from the theory of science whereas he was the true experimentalist in the best sense. Honesty and completeness of statements of facts were his *forte*. We unfortunately never had occasion to meet, which I deeply regret. He wrote on p. 260 that I had misquoted some of his conclusions in *Popper versus Einstein* but since the only direct quotation in my book was *verbatim* (p.193) I think he must have meant that I had misunderstood him.

196   Presenting it like that is common to contemporary German textbooks on special relativity as, for instance, Roman Sexl/Herbert K.Schmidt, *Raum – Zeit – Relativität*, p. 39 f.; W. Greiner/J. Rafelski, *Spezielle Relativitätstheorie*, p. 25 f; Ulrich Schröder, *Spezielle Relativitätstheorie*, p. 41f. The textbook by A. P. French, *Special Relativity* (ed. 1997), p. 101–104, does not mention the Hafele/Keating-experiment but the logical mistake is the same in principle because French also confuses clocks and time. I did not look for equivalents in other English textbooks.

197   The scientists at the CERN recently took that attitude again by terminating the neutrino experiments mentioned in the *Introduction* as soon as they had shown sublumi-

a scientific theory empirically if we determine only afterwards what it predicts?

I must leave those questions unanswered here for the moment. Even now it ought to be clear that scientific research addressing itself to 'basic laws' only, that is to ultimate and indisputable scientific truths, will hamper the progress of science itself. It encourages the tendency to obstruct new discoveries. I remind once more of Einstein's words:

> 'The great discoveries give me only little pleasure because they do not seem to me, at present, to make the understanding of the foundation any easier.'[198]

Mathematical deduction leaves error no place. Understanding science as based on deduction alone can explain error only by personal deficiency or by dishonesty. No one sharing that belief will understand that it is not a sign of weakness but on the contrary a highly commendable virtue if an *empirical* theory exposes itself to the risk of being refuted by experiment. Theoretical physics will make progress only if it recognizes *in theory* the possibility of error. It must learn to accept that our knowledge of the empirical laws of nature does not come from deduction but from speculation, which we can only *criticize* by observing the rules of logic and mathematics and which we can only *test* by experience. It must start from an understanding that does not take the boundless knowledge of man as its starting point but on the contrary his boundless ignorance.

---

nal velocities. I will discuss that in more detail at the beginning of *Chapter 9.*

198 See *sec. II, 2*, above.

SECOND PART:
A CENTURY LOST – THE LONG RETURN TO ETHER THEORY

,Although I am deeply in love with the waves (much
more than with the particles) I nevertheless believe
that the quantum theory is in a very definite sense
a particle theory (here I disagree with Schrödinger)
and in a sense which excludes a duality, or analogy,
or complementarity, between particles and waves.'

KARL POPPER

Of the various approaches to the theory of science, only the deductive approach remains in empirical science; its use not for generating but only for criticizing empirical theories. II. Returning to ether theory overall topic of Second Part; deductive theory can serve for practical application and for criticizing theories; remote deductions more interesting for scientific research than events of everyday life; criticizing quantum theory and the theory of relativity inevitable for explaining gravity; success of quantum theory and the theory of relativity partly due to ambitions of Prussian government and partly to psychological mechanism; the latter causing belief in 'community' of scientists; criticism only of theories; personal criticism not intended. III. Distinction between 'descriptions' and 'explanations' important for Second Part.

The *First Part* of this essay showed fundamental mistakes in Einstein's approach to the problems of the theory of scientific knowledge. His axiomatic method, which he believed to apply not only to geometry but also to physics, consisted in fact in patching together those two without heeding the difference between them. He must have believed the rules of geometry to be laws of nature itself. Only that would explain why he did not distinguish clearly between the empirical science of physics and

157

the non-empirical discipline of geometry[199]. That view would have been mistaken however.

Physics has to do with reality, and geometry may also have to do with reality. But whereas the object of physics is reality itself, geometry has no such object[200]. It is an instrument which we can use for describing real things or situations but also for describing things or situations not existing in reality such as for instance constant motion on straight lines. History clearly shows that geometry is a creation of the human mind. The ancient Egyptians probably invented it for redistributing fertile land after the annual inundations of the Nile. At any rate it is part of mathematics, belonging to *language* as we saw in *Chapter 5, II*.

Physics, by contrast, is about aspects of reality that are independent of man. Keeping apart physics and geometry is important therefore, but Einstein seems not to have been aware of that. At the beginning of the 20th Century he would not have been alone with that as Poincaré showed[201]. Confusing mathematics and physics was quite common then. That is why Einstein's axiomatic approach to the theory of scientific knowledge could find followers so easily although he had based it on a misunderstanding.

Of the other approaches to the problems of the theory of scientific knowledge the inductive approach discussed in *Chapter 3* probably still is the most popular in theory. Physicists often cling to it because they trust in Newton's

---

199 In *Popper versus Einstein* (p. 196–219), I have discussed that distinction in more detail, and explained some of the reasons giving rise to the belief that mathematics could be prescribed by nature itself. The overwhelming success of Newtonian theory probably was strongest, but the mathematical regularity of the periodic system must also have been important. - I think the best counter-argument against that belief is that we could translate into normal language any expression made in mathematical symbols in any section of mathematics including geometry. That would make those expressions much longer and far more difficult to understand; but it would not affect their meaning if correctly done. That shows that those expressions are manmade, and so is everything we derive from them.

200 My criticism particularly concerns Einstein's concept of ‚space'. I will explain it in *Chapter 11, II*.

201 Poincaré was one of the few to see the problem clearly, for instance in *La Science et l'Hypothèse* (1902) p. 51–57.

authority and therefore believe doubting inductivism to be unrasonable[202]. Yet that approach also proved to be mistaken as we saw there. The empirical assumption on which it relies is not true and the inferences which its adherents draw from it consist in logical fallacies. In spite of Newton's belief in induction no mathematician or physicist ever succeeded in establishing a valid rule of inference that would permit inferring from observations made in the past statements on events that are to happen in future.

I

The only rational approach to the problems of scientific knowledge remaining after that is the *deductive* approach mentioned in *Chapter 3, I, 1*. It leads only to a limited rationality however, because it does not show a way of *generating* knowledge but only a way of *criticizing* it. Popper convincingly showed that in his *Logik der Forschung*[203].

Generating knowledge is a rational process only partly, even in empirical science. A scientist must be aware of problems and try to find solutions to them. Being aware of problems is something most people can learn. But finding solutions to them depends on the gifts of imagination and intuition. If we face an empirical problem of science which no one ever solved, then we must *invent* a solution to it because there is no other way of finding one, not even in physics. And inventing solutions has nothing to do with deduction. The deductive aspect comes in only *after* we have found a tentative solution, when we or others criticize it by analyzing its implications and by comparing them with reality or with other theories. We saw that in *Chapter 5, II*.

The laws of nature on which we base our inferences are in the way of hypotheses because nothing can prove their truth whereas a single exper-

---

202 For Newton's belief in inductivism see the quotation from his *Principles* in *Chapter 2, I,* above.

203 Popper, *Logik der Forschung*, p. 2–21 (*The Logic of Scientific Discovery*, p. 21–48).

iment would refute them if it could be repeated. Some may dislike that because it means that in the field of empirical science there can be no certain knowledge at all. But if hypothetical knowledge is all we can get, then facing that fact is more veracious than wishfully pretending to knowledge not existing. Popper's critically deductive approach thus recognizes the limits of rationality itself as well as the fact that the attitude of the true scientist must be founded in ethics. In this *Second Part* I will therefore rely on the *critical deductive* approach in the sense of Popper's theory.

The overall topic of this *Second Part* will be that we must return to *ether theory* if the theory of light and matter is ever to make progress again. Resuming some variety of the ether hypothesis or proposing a new one if necessary is the only way left open for explaining gravity which at present is the greatest open problem of theoretical physics. It is also my principal reason for writing this essay. We will see in *Chapters 11* and *12* that a new variety of the ether hypothesis can explain not only gravity itself but also several other physical phenomena, and that it can open approaches to explaining many more phenomena remaining unexplained at present.

II

My greatest problem in this *Second Part* will be how to make not only outsiders but also theoretical physicists realise the necessity of falling back on ether theory. Giving it up was a matter of fashion at the beginning of the 20th Century, just as believing in Big Bang theory is a matter of fashion in our time. Searching for mathematical solutions to empirical problems agreed with the intellectual climate of that period although there never was reason for believing it to be the correct approach[204]. Empirical problems can only be solved by empirical explanations. Returning to ether theory or to some other empirical hypothesis is necessary therefore if theoretical

---

204 See the quotation in *Chapter 3, III, 4* from Ernest Nagel/James R. Newman, *Gödel's Proof.*

physics is ever to get out of its present deadlock. But it will hardly succeed as long as a majority of theoretical physicists believe quantum theory and the theory of relativity to provide correct descriptions or even explanations of physical phenomena. All those living in our time must have grown up in the axiomatic tradition of science prevailing in the 20th Century. To them, falling back on ether theory would mean to them giving up much of what they learned in their academic education. And to some it would also mean giving up what they taught others to believe.

One of the effects of that situation seems to be that theoretical physicists tend to ignore criticism coming from outside. At least I myself never had the good fortune of meeting one who would have discussed to the end with me the questions I had. Several theoretical physicists were at first quite willing to explain to me the theory of relativity. Corresponences usually began very kindly. When I raised counter-arguments however, they always ended unilaterally from the other side when they were just about to get interesting. That happened to me more than once. Arguments can take no effect on someone unwilling to read them or to listen to them. I never succeeded in overcoming that attitude of theoretical physicists and have long been wondering what might have caused it. I could offer several explanations now but will mention only two of them here, both merely conjectural.

A political constellation now lying in distant past may have contributed to bringing about the present unsatisfactory situation. The Holy Roman Empire had collapsed in 1806 in the Napoleonic wars and the Second German Empire, much smaller than the first, was founded at Versailles in 1871. Berlin became its capital and from then on attracted more attention than ever before. Prussian ambition may have required that it became the capital also of German science. It needed publicity for achieving that, and what way of attracting it could have been better than by advertising the merits of scientists working there? At any rate public interest in the activities of scientists working at Berlin seems to have taken a steep rise in the last decades of the 19th Century[205].

---

205 Armin Hermann, *Max Planck* in *Die Großen Physiker*, ed. by Karl v. Meÿenn

The fact that quantum theory and the theory of relativity found almost general acceptance so quickly might have been the effect of that publicity instead of being its cause. The Nobel Prizes awarded to Max Planck in 1919 and to Einstein in 1922 would then have reinforced that effect, and the same would apply to Einstein's becoming a member of the British Royal Society in 1919 and to his being called to Princeton in 1932. Going into details of that development would be beyond the scope of this essay. At any rate it resulted in a situation in which returning to ether theory would have meant losses of prestige not only for the two scientists themselves but for many others as well.

Seen from the point of view of science Max Planck and Albert Einstein who seemed to be the intellectual heroes of their time may well have been its victims thus. They gained general recognition and probably were content with the popularity that conveyed on them. But if we try to imagine what ways would still have been open to them if they themselves had discovered the mistakes they had made, then the situation would look quite different. We will see that in the following chapters. Planck would then have had to admit that there neither was any reason for assuming his quantum of energy h to be indivisible nor for believing it to be a constant of nature. Einstein would have had to confess that he had misunderstood Planck's theory when he proposed his own light quantum hypothesis, that he had based his special theory of relativity on a self-contradiction and that there never was any empirical reason for believing the speed of light to be constant. Admitting all that would have been disastrous for the reputation of both of them.

The other reason which might explain the tradition in theoretical physics of not taking notice of criticism coming from outside is at the level of psychology. Understanding quantum theory and the theory of relativity is difficult for anyone. That is a natural consequence of their logical inconsistencies which we will see in the following chapters. The more difficulties some theoretical physicist once experienced with understanding

---

(1997), vol. II p. 143, 145. Max Planck also reports that in *Zur Geschichte der Auffindung des Wirkungsquantums* (1933) in *Vorträge und Erinnerungen*, 5th ed. (1949), p. 15, 21.

those theories himself, and the more he believes to have overcome those difficulties, the less will he be inclined to take notice of anything a mere outsider may say against the theories. Avoiding waste of time need not be his only reason for that attitude. He may also be afraid that someone might come up with an argument he cannot meet. And he may have heard that other theoretical physicists have similar problems with those stupid outsiders who keep pestering them with requests for explanations without being able to understand the answers given them. In the course of time those effects will teach him to take no notice of criticism coming from outsiders.

I consider that attitude particularly dangerous because it has a tendency of reinforcing itself. The symptoms of that tendency show in theoretical physics of our time. The more complicated physical theories get, the less will any physicist be inclined to explain them to an outsider, let alone discuss them with one who is critical of them. At the same time the very complication of those theories is likely to encourage in any physicist believing to have understood them the feeling of being one of the elect. When I learned some years ago that theoretical physicists consider themselves as belonging to a *scientific community* or to a *fraternity,* I was taken aback because that kind of sectarianism does not exist in the legal profession nor to my knowledge among historians or economists or in any other academic discipline. Theoretical physicists however seem to feel themselves as belonging to a chosen community, and I doubt whether they would admit to it even chemists or biologists.

Quantum theory and the theory of relativity had a considerable share in creating that situation. I cannot avoid criticising those theories before coming to the explanation of gravity therefore. I will begin by discussing quantum theory, the theory of relativity and the planetary model of the atom, and by applying to those theories the principles explained in the *First Part.* My criticism will not aim at persons but only at the theories in which they believed. Only after having shown the insufficiencies of those theories can I discuss the problem of gravity and propose a way of explaining its physical effects by a new variety of the ether hypothesis.

III

Distinguishing between 'descriptions' and 'explanations' as shown in the *Introduction* will be particularly important in the following chapters. Many physical phenomena remain unexplained at present. Only that distinction will show that assuming an otherwise unknown substance to be filling the universe is the only way left open for finding explanations to them. There is no other way of introducing new *empirical* information explaining 'the known by the unknown'.

Ether theory will be an important aspect for its own sake in this *Second Part* therefore, but even more so for evaluating other physical theories and for preparing the ground for the explanation of gravity which I will propose in *Chapter 11*. But we must begin by trying to understand the present problem situation, and that means that we must begin by looking at its history.

CHAPTER 7:

ON THE STATUS OF ETHER THEORY IN THE 19ᵀᴴ CENTURY

History of ether theory. I. Descartes three elements; the rationality of his approach; his introducing the principle of locality (as opposed to action at a distance); Rœmer's discovery of the speed of light; the problem of transportation of light; Newton's particle theory; Huygens' wave theory in analogy to sound; transportation of energy through elasticity of ether; wave theory prevails; it explains also other phenomena; finite speed of light raises problems for mathematics; speed in outer space unknown; ether theory weakened by doubts: longitudinal or transversal waves; variety of theories: ether stationary (ether wind) or dragged along (Fresnel). II. Experiments: Fizeau's water experiment; interpreting Michelson's experiment; explanation of white starlight; spectral analysis; no mention of analogy of light and sound in modern textbooks; experiments refuted stationary ether only. III. Ether theory's power of explanation, approaches also for other phenomena.

Ether theory was probably first conceived in ancient Greece. Hesiod already mentioned ether and to Aristotle it was the divine substance enclosing the earth. Along with fire, water, earth and air it was the fifth element, the *quintessence* to which he attributed among other things also geometrical properties[206]. About in that vague shape the ether hypothesis was incorporated also in Ptolemy's theory of the spheres in the 2ⁿᵈ Century. And that was where matters rested without major changes for centuries to come.

---

206   Aristoteles, *De Caelo*, Book I, Chapter 1, 3.

I

Things gradually got going again in the 17th Century when ether was re-invented as it were. Copernican theory had overcome Ptolemaic theory and had created havoc in many frozen systems of thought. In that difficult situation ether theory owed its miraculous resurrection mainly to the great French mathematician René Descartes (1596–1650). He was the first philosopher of modern times to develop a theory that was to explain matter and all other physical phenomena by only three elements, each of them consisting of tiny spherical particles standing in direct contact with one another[207]. The smaller particles of the finer elements would fill the gaps between the larger particles left open by their spherical shape. Direct contact of those particles was to explain by strictly mechanistic principles all physical phenomena known in his time, and his *second élément*, an invisible substance which he put in a class with fire, was to fill all space.

Some may smile at Cartesian theory in our times but whatever one may think of its results: in the *method* of science it certainly was a remarkable improvement. Descartes made the first important step towards overcoming Aristotelian doctrine which in almost two millennia had been content with accepting the 'essence' or the 'nature' of things also as their explanation. Aristotle explained gravity for instance by claiming that heavy elements have a 'natural tendency' of approaching as near as possible to the centre of the cosmos which to him was the centre of the earth, whereas light has a 'natural tendency' remove from it. That 'explanation' presupposed as given all that was in need of being explained, the tendency of approaching or removing as well as its being natural. It gave no empirical information but simply put an end to discussion by feigning an insight not existing. Notions about the nature or the essence of things can never serve as explanations because they may differ from person to person. They cannot replace missing arguments[208].

---

207 Descartes, *Traité de la Lumière* (1664), particularly Ch. 5 (vol. XI p. 1 ff.).
208 For a detailed discussion of that view see Popper's *Open Society*, Ch.11 section II

Descartes' theory of three elements, by contrast, gave to human mind a host of new information, albeit in the shape of a highly speculative hypothesis about the existence of invisible substances, but permitting also to criticize that hypothesis at the hands of reality. In his approach any explanation would extend human knowledge, if only by introducing new hypotheses. It would also raise new questions however, which would then in turn be in need of explanations. Thus it would not only extend our knowledge but also our consciousness of our own ignorance which is the condition of curiosity and therefore of all progress of science. Despite the mistakes Descartes may have made, his *method* was in fact the beginning of modern science and epistemology[209]. It created the intellectual climate needed for rational discussion, and his way of approaching the problems of physics justly dominates the ideas of physicists to this day.

Descartes had based his theory of three elements on the notion that explaining physical processes must consist in describing the behaviour of physical bodies standing in direct contact with one another. This mechanistic principle, now also named *principle of locality* (as opposed to *action at a distance*), is widely recognized still today. Some may call it into question but as a principle of methodology it remains an indispensable part of any explanation deserving the epithet 'physical'[210].

(*1*) Descartes was in a different situation than later generations however, because he did not know that the speed of light is finite. We can hardly appreciate the merits of his theory without taking that into account. He still believed light to propagate instantaneously and tried to visualize that by imagining light rays to be like rods ('batons') consisting of a rigid

---

(vol. 2, p. 9–21).

209 That must also have been Huygens' view. According to him, Descartes treated matters of physics better than anyone had treated them before him (Christiaan Huygens, *Traité de la Lumière* [1690], p.7 f.). Descartes even proposed a rational approach to explaining gravity to which I will return in *Chapter 11, IV.*

210 The main reason for that is in methodological nominalism. I will explain it in more detail in *Chapter 11, III.*

sequence of particles in which the motion of one end would be transmitted to the other end by physical contact without delay[211].

That explanation became untenable only after Descartes' death (1650) when the Danish astronomer Ole Rœmer (1644–1710) discovered that the speed of light is finite (1676). His discovery was a remarkable feat of physical imagination. The periodic eclipses of Jupiter's moons had long been observed but Rœmer seems to have been the first to notice that the time intervals between them varied in periods of six months. He surmised, brilliantly I think, that the cause of those fluctuations could not be in the orbits of Jupiter's moons because their periods were quite different, but must have to do somehow with *the earth's* orbiting the sun. His next surmise was that light was taking more time when Jupiter was at greater distance from the earth and less time when at shorter distance, which implied that the duration of a light signal depended on the distance it travelled[212]. Closer investigation confirmed that daring conjecture. Since then science had to face the fact that light will take *time* for its journey through space. The question of its propagation assumed new importance and gave rise to new speculations.

If light takes time for travelling from its source to the objects it illuminates, then the inference seemed inevitable that light radiation is a process of transport. That raised the questions of what was being transported and how it was being transported. Was it a transport of something material or a mere wave undulation which, though also requiring a transporting medium, need nevertheless not consist in transporting some object? That seemed to be the crucial question.

Christiaan Huygens, the great Dutch physicist, astronomer and inventor, could imagine only as an oscillation of elastic particles a velocity so high as that of light[213]. In his *Traité de la Lumière* (1690) he surmised that some invisible substance which he called 'matière éthérée' or simply *éther*

---

211   Descartes, *Traité de la Lumière* (1664); Ch. 5 (vol. XI p.98 ff.).

212   Huygens gave a detailed descripiton of Rœmer's discovery in his *Traité de la Lumière* [1690], p. 8 f.

213   Huygens, *Traité de la Lumière*, p. 14-26.

must be filling the whole universe. He pondered deeply on the properties it would have to own if its existence were to agree with the visible manifestations of light[214]. Like Descartes he assumed ether waves to be consisting of tiny particles packed so tightly that they were always touching one another. He did not think they had to be in the shape of balls however, and expressed more clearly than Descartes had that if they were to permit any motion at all, they would have to be highly elastic[215].

Starting from an analogy to the behaviour of sound his notion of smallest elastic ether particles enabled Huygens to explain in a highly speculative theory and in strict compliance with the principle of locality, which he considered very important, the phenomena of light as direct actions of ether particles on adjacent particles[216]. Due to their high elasticity and their immediate contact with one another, light would propagate in spherical waves analogous to the propagation of sound. The high elasticity of ether particles would cause extremely quick, indeed by human standards almost instantaneous reactions to impulses coming from adjacent particles. And the propagation of light would consist only in those reactions while the particles themselves kept their places[217]. Based on that surmise he conceived plausible explanations for most of the phenomena of light known in his time. Huygens thus became the founder of the *wave theory* of light[218].

Newton also considered that theory but he had taken a different approach already in his famous *New Theory About Light and Colors* published

---

214   Huygens, *Traité de la Lumière*, p. 12.

215   Huygens, *Traité de la Lumière*, p.16, 17.

216   For the analogy to sound see Huygens, *Traité de la Lumière*, p.4. - At the beginning of that treatise (p. 3) Huygens mentioned 'la vraie philosophie, dans laquelle on conçoit la cause de tous les effets naturelles par des raisons de mécanique. Ce qu'il faut faire à mon avis, ou bien renoncer à toute espérance de ne jamais rien comprendre dans la physique'. ('True philosophy, in which one conceives the cause of all natural effects by mechanic reasons; which, in my opinion, is what must be done, or give up all hope of ever understanding anything in physics.' - My translation).

217   Huygens, *Traité de la Lumière*, p. 14–18.

218   Volkmar Schüller, *Christian Huygens*, in *Die Großen Physiker*, vol. I, p. 194, 208.

in 1671[219]. In that paper he had investigated from the experimental side the questions raised by the phenomena of refraction, and had discussed the possibility of light rays to be consisting of *particles*[220]. Later, in his *Hypothesis explaining the properties of light* (1675), he explained that approach in more detail[221]. Science therefore considers him the author of the particle theory of light which some less clearly now name 'emission theory'[222]. There will be more about his approach in *Chapter 9, VI, 2*.

At the beginning of the 19th Century speculations about the exixtence of ether found new support. The phenomena of interference, demonstrated in 1802 by the British physician and physicist Thomas Young (1773–1829) in his famous double slit experiment, and the polarization of light discovered by the French physicist Etienne Louis Malus (1775–1812) in 1808 seemed to indicate almost inevitably that light must be an effect of waves. That led to the hypothesis of transversal light waves proposed by Young which the French physicist Augustin Jean Fresnel (1788–1827), around 1815–1821, expanded into a comprehensive theory of light including all the phenomena known at that time such as interference and polarization, refraction, birefringence and aberration.

(2) New problems also arose from those discoveries however. One of them was at the level of mathematics and resulted from the finite speed of light. Rœmer's discovery had not only expanded human knowledge but had also faced physical theory with new problems. Until then, light had always been the means for measuring other effects. Now it had come in need of being measured itself, which thoroughly upset all mathematical approaches.

---

219 Isaac Newton, *A New Theory About Light and Colors*, Philosophical Transactions 6 (1671), p. 3075–3087.

220 In fact, Newton discussed the possibility of light rays to be consisting of particles but did not take sides on the issue (Newton *loc.cit.* p. 3078; my italics).

221 It was published by Thomas Birch in *The Hisory of the Royal Society*, vol. 3(1757), p. 247–305.

222 Ivo Schneider *Isaac Newton*, in *Die Großen Physiker*, vol. I, p. 185, 191 f. - Distinguishing Huygens' theory from Newton's by using the terms 'wave theory' and 'particle theory' is clearer than by the terms 'wave theory' and 'emission theory' because Huygens' theory also assumes emission, though not of particles but of energy.

If the speed of light were infinite, then light signals could serve for calculating distances without any problem. Dispatch and arrival would be simultaneous. For purely theoretical reasons no time interval would be possible. The discovery that the speed of light is finite implied that time will pass between dispatch and arrival of a light signal. Due to its high speed the interval would be negligible at close range. Where great distances were in question however, such as those between celestial bodies, any calculation aiming at precision would have to allow for light signals taking time on their way, possibly even much time. In astronomy and cosmology we often speak of light years after all, sometimes even of millions of light years. If both our earth and the celestial bodies we observe are moving while light is on its way, then we would need mathematical formulae allowing for that motion for calculating the positions of other celestial bodies at the time of arrival of the light signal here on earth.

The discovery of the finite speed of light thus made algebraic approaches to cosmology far more complicated than they had previously been, and that was not nearly all. Another problem, apparently unforeseen by many physicists, arose at the level of empirical knowledge. How could we calculate the distances and velocities of celestial bodies at all without knowing exactly at what velocity light travelled? Clearly, their changing positions and the time intervals lying between them would have to be the starting point, but observing them would have to depend on light signals. Any calculation of distances would have to allow also for the duration of those signals.

What size were we to insert for the speed of light? Various terrestrial measurements had been made after Rœmer's discovery [223]. Could we be sure however that the speed of light measured here on earth is the same as that in outer space? Sound for instance travels much faster in water than in air. Could the speed of light not also vary? How could we find out more about that without going out into the universe ourselves? Yet even if we got there, which seemed inconceivable at that time, we would still need light signals for measuring. One question led to the next, and more

---

223   For details, see Bernard Jaffe, *Michelson and the Speed of Light* (1971).

unknown operands turned up each time than equations for determining them. Mathematical astronomy seemed to be at its end.

(3) The discoveries of interference and polarization also raised new problems in the theory of light. They concerned ether theory and were at the level of experimental physics.

The mechanistic interpretation of light in analogy to sound proposed by Huygens seemed to imply that light will propagate through impulses in the direction of propagation, hence in longitudinal waves. Huygens' notion had been that by virtue of its elasticity every ether particle would pass on to the next particle an impulse received. The observations on interference and polarization made by Young (1802) and Malus (1808) and Fresnel's theory mentioned above seemed to indicate however that light waves must consis of transversal waves, behaving more like the vibrations of the strings of a violin.

That problem was also unresolved still when Einstein began his academic career. It did not affect ether theory itself initially because transversal light waves also seemed inconceivable without some medium. If there was oscillation then something must oscillate, so one thought, and that seemed to justify the conclusion that it must exist. Most physicists were convinced of the existence of an invisible substance in which oscillations take place, and gave it the name 'ether'. Throughout the 19th Century hardly any physicist seems to have doubted the existence of ether. Yet the discoveries of interference and polarization and the question of the waveform they raised had not only given ether theory new support but had also deprived it of some of its power of persuasion. Admittedly it could explain much but apparently not everything. Instead, it seemed to raise problems itself. The two basic beliefs of theoretical physics of that century, that in the omnipotence of mathematics and that in the existence of an invisible substance called 'ether', seemed to conflict. Having different opinions became possible. One could even dare to contradict one of them.

(4) The fact that its supporters did not agree among themselves further weakened ether theory. At the end of the 19th Century there existed not just one ether hypothesis but several varieties competing with one another.

One of them assumed ether to be 'at rest' and filling the whole universe. For bringing that in line with the visible movements of the planets we would have to assume the earth to be in motion relative to ether. Ether would have been stationary then, and matter moving through it would have been permanently exposed to an 'ether wind'. By terrestrial standards it would in fact have needed to be a violent hurricane. Relative to the sun the earth is travelling through space at a speed of about 30 *km/s* ($\approx$ 62,000 *mph/s)* and her speed relative to more distant celestial bodies would have to be even higher. Assuming ether to be stationary it therefore seemed worthwhile to try making the motion of the earth relative to ether visible in experiments.

Against that stood another hypothesis conceived by Fresnel in 1818 already[224]. For making plausible his theory of transversal waves he had surmised that ether would adapt its velocity to that of matter and be 'dragged along' by it as it were. If that were true, then no ether wind was to be expected because any motion of matter would be the same as that of the ether surrounding it. Every single celestial body would be in a field of ether stationary relative to that body, but within ether itself there would be movements or currents comparable to those of the clouds in the sky.

Those varieties of the ether hypothesis naturally led to different approaches in the attempts at making the effects of ether visible in experiments. Various experiments were made but they turned out to be uncommonly problematic.

On the one hand difficulties arose from the fact that light is the smallest physical effect to which our sensory organs will respond in a way that we can notice directly. If it consisted of waves and if ether was the medium transporting them, then the components of ether, if existing at all, would be smaller even than those of light, just as water molecules for instance are smaller than the waves of water. From that alone it followed that our senses would never be able to perceive the components of ether directly. The only question was whether it would be possible to make the motion of ether indirectly visible through other effects.

---

224 A. P. French, *Special Relativity*, p. 45.

On the other hand that immediately led to the next question of how anyone could make such indirect effects visible at all without knowing more about them. What kind of effects were we to expect from a fabric we had invented only for explaining other physical phenomena? What kind of experiments were to demonstrate its existence? One did not even know after all whether ether was at rest or moving.

Seen from that angle Fresnel's hypothesis of *moving* ether must have aroused particular misgivings because so much seemed to hinge on the question of whether or not ether is at rest. Newton's first *Hypothesis* in his *System of the World* had been

'that the centre of the system of the world is immovable[225].

Many therefore believed Newtonian theory to apply only to a state of absolute rest[226]. From their point of view the existence of moving ether would have meant that we had to abandon Newtonian physics altogether. Besides, if ether were being 'dragged along' by matter, then it would have evaded empirical observation more than anything ever had done before. What physical effects were we to expect at all from an invisible substance that was not in motion relative to matter? How could we make it visible at least indirectly? Did not the only hope of man ever being able to demonstrate the existence of ether waves lie in the assumption that there must be an ether wind? By not answering those questions and by leaving unresolved the problem of longitudinal or transversal waves the hypothesis of moving ether seemed to fall between two stools. In experimental

---

225 Cajori, *Sir Isaac Newton's Mathematical Principles of Natural Philosophy and his System of the World*, vol. 2, p. 419.

226 Newton himself seems to have considered the question irrelevant. At the beginning of Section XI of his *Principia* he wrote: 'I have hitherto been treating of the attraction of bodies toward an immoveable centre; *though very probably there is no such thing existent in nature*' (Cajori, vol.1, p.164; my italics). I believe that view to be correct but Einstein shared the belief mentioned in my text. The first four pages of his paper *Zur Elektrodynamik bewegter Körper* (On the Electrodynamics of Moving Bodies) show that. He did not distinguish clearly between reality and systems of coordinates. We saw that in *Chapter 5, III*, already and will see it in more detail *Chapters 7* and *8*.

physics it was notorious for contradicting itself and deliberately evading experimental examination. And theoretical physicists took umbrage at its speculative nature and its ostensive conflict with Newtonian theory. As a result, neither of the camps took Fresnel's hypothesis of moving ether seriously although their reasons for that were contradictory.

II

Theoretical physicists thus faced a highly complex situation in the 19th Century. Many dubious assumptions were interwoven, and understanding the issues depended on distinguishing clearly between observations and interpreting theories, and on evaluating them critically. However, while they were struggling with those problems experimental physicists were not idle.

(*1*) As mentioned, Fresnel had surmised in 1818 already that ether itself might be in motion and be dragged along by matter. In 1851 the French physicist Hippolyte Fizeau (1819–1896) thereupon designed an interesting experiment for testing Fresnel's theory. A beam of light was split in a semi-reflecting mirror. The parts propagated through flowing water in opposite directions, then reunited and were measured for interferences (see *Figure 2 on the next page*).

The result left no room for doubt. There were interferences, and they depended on the velocity of the flowing water. Light was being 'dragged along' by it to some extent.

In the 19th Century most physicists considered that as conclusively confirming the hypothesis of moving ether. Discussions were mainly about whether similar observations were possible also without water. Would the movement of matter in space also influence the speed of light? That was what the American physicist Albert Michelson wanted to find out in his famous experiments made in 1881 and 1887, the latter together with his compatriot Edward Morley. The outcome is well known. In spite of utmost diligence in the execution of the experiments, considering and eliminating every conceivable source of error, and in spite of several repetitions there never was a positive result. They could not detect any motion of the earth relative to ether.

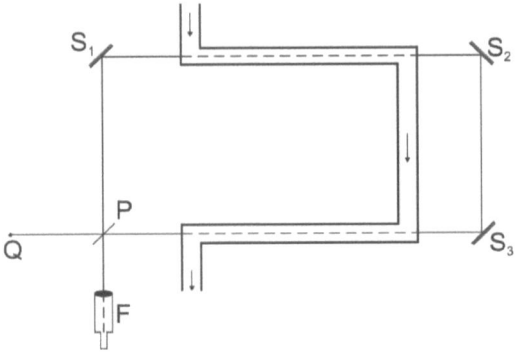

Figure 2, Fizeau's experiment, schematic diagram: A beam of light emanating from source F is split by semi-reflection in P. Its parts traverse flowing water in opposite directions, re-unite again through semi-reflection in P and are measured for interferences in Q. [227]

Michelson himself interpreted that as showing that his experiments had refuted the hypothesis of *stationary* ether[228]. He never commented on other varieties, at least not to my knowledge. In fact, his experiments did not get Fresnel's theory of *moving* ether in any difficulties at all. They could even be interpreted as endorsing it. If the atmosphere of the earth 'drags along' ether, then an ether wind directly observable in those experiments could not reasonably be expected from the outset. Few seem to realise that in our days[229].

227   Quoted from Max Born, *Die Relativitätstheorie Einsteins*, p. 120.

228   Jaffe in *Michelson and the Speed of Light*, p. 76, quotes Michelson as having said 'The hypothesis of a *stationary* ether is erroneous.' (Jaffe's italics!).

229   Peter Day, *The Philosopher's Tree, A Selection of Michael Faraday's Writings*, p. 103 is symptomatic. He considers Michelson's experiment as having refuted the ether hypothesis altogether.

(*2*) Einstein may not have been acquainted with Michelson's experiments when conceiving his special theory of relativity; at least he did not mention it in that paper[230]. But other important considerations also pointed in the direction of Fresnel's hypothesis of ether being 'dragged along' by matter, particularly some interpretations of the phenomena of light. Physicists already knew them at the end of the 19th Century. Hence they too were part of the background before which we must see Einstein's theories, and we must take a look at them.

(*a*) Observations indicate that relative to the earth other celestial bodies move at very high velocities. The speed of the earth on its orbit around the sun is about 30 km/s. In distant stars or galaxies we must assume much higher relative velocities because many of them are much larger than our Milky Way and rotating around their own centres millions of light years away from the earth. Long before Einstein that raised the question of what influence those velocities might have on light reaching us from other celestial bodies.

Generally the addition theorem will apply to mathematical representations of the motion of physical objects. Depending on direction we will add to or subtract from one another their relative velocities, using the vector calculus if necessary. For motion on straight lines Einstein demonstrated that in his famous railroad example[231]. If a traveller in a moving train moves in the direction of its motion, then his speed over ground will be calculated by adding his own speed within the train to that of the train itself, and if he moves against that direction it will result from the subtraction of those two velocities.

If the same principle applied also to light, then light coming to us from distant celestial bodies would reach us at different velocities. According to the Doppler principle we would expect the motion of the light source to

230 Albert Einstein, *Zur Elektrodynamik bewegter Körper* (On the Electrodynamics of Moving Bodies), Annalen der Physik vol. 17 (1905), p. 891 ff.

231 Albert Einstein, *Über die spezielle und die allgemeine Relativitätstheorie* (On the Special and General Theory of Relativity, 1917), p. 6 ff.

cause an increase or decrease in the frequencies of light itself. Due to the high differences of relative velocities which we must assume in distant celestial bodies we should therefore expect significant shifts of frequencies. First thoughts seem so to suggest therefore that the stars in the night sky should have different colours. Anyone can see with their own eyes however that this is not so in fact.

In the first half of the 19th Century physicists could still believe they knew a plausible explanation for the fact that all stars radiate the same white light. They could assume that the Doppler effect caused by differences in relative speed would affect not only the visible spectrum of light but also its invisible ranges. An increase of frequencies would then shift the high frequencies from the visible range to the invisible ultraviolet range while at the same time shifting the previously invisible lower frequencies from the infrared range into the visible range. As a result the human eye would always see the full visible spectrum of white light. Not the physical properties of light would then explain the white light of the stars, but the limits of human visual perception. As long as physicists assumed that notion to be true they could use the addition theorem of mathematics without any problem also for calcuating light.

However, that explanation became untenable in 1859/1860 when Robert Bunsen (1811–1899) and Robert Kirchhoff (1824–1887), both of Heidelberg University, had invented spectral analysis which permitted an accurate determination of the frequency shift also in white light by observing its spectral lines[232]. Their method not only revealed that the spectrum of light coming from distant stars corresponds to that of our sun but also showed that there was no significant shift of frequencies. The spectral lines had shifted indeed, but by far not to an extent that would have correlated with the speed differences which we must assume. Under no circumstances could the observable shift plausibly explain speed differences of 30 km/s or more. The inference seemed inevitable that the proper

---

232 The spectrum of light emanating from some gas, as for instance hydrogen, is not continuous but shows a pattern of blanks characteristic of that specific gas. These blanks are the 'spectral lines' of its spectrum.

speed of the celestial bodies from which light came did not significantly affect its speed. Apparently addition or subtraction of velocities did not accurately describe the propagation of light. That probably was the origin of Einstein's assumption that the speed of light in a vacuum will always be constant, independent of the motion of its source[233].

(*b*) Seen from the angle of ether theory however, the observation clearly spoke in favour of wave theory once again. Fresnel's hypothesis assuming ether to be 'dragged along' by matter could explain it quite naturally by an analogy to sound, therefore by a parallelism to a physical phenomenon which everyone knows and to which light compares in many respects.

The analogy to sound is important for understanding the different approaches to the problems of explaining light. Nevertheless it seems to be almost forgotten in our time. Modern textbooks on the theory of relativity do not mention it any more although it almost forces itself upon readers more interested in physical explanations than in mathematical descriptions[234]. At the turn of the 19th Century it must still have been remembered because Huygens had already seen it[235] and Faraday had explicitly pointed to it when he wrote in 1857:

> 'The nature of sound, and its dependence on a medium, we think we understand pretty well. The nature of light as dependent on a medium is now very largely accepted.'[236]

The analogy deserves closer consideration therefore.

---

233 I presume that one of Einstein's teachers at the Polytechnikum at Zurich may have inserted the speed of light as a constant in his equations. For practical purposes that would have been reasonable because it may be as near to the truth as we can get (see *Chapter 12, II*).

234 For my definition of the terms 'description' and 'explanation' for the purpose of this essay see section *II, 1* of the *Introduction,* above.

235 Huygens, *Traité de la Lumière* (1690), p. 4; Schüller, *Christian Huygens,* in *Die Großen Physiker,* vol. I, p. 185, 191 f.

236 Quoted from Peter Day, *The Philosopher's Tree, A Selection of Michael Faraday's Writings,* p. 104.

Given identical frequencies at their source, the frequencies of sound reaching our ear from a source approaching us at high speed will be higher than those coming from a stationary source, and they will be lower if the source is moving away from us. In our times of fast moving vehicles everyone will be acquainted with that phenomenon called 'Doppler effect'. And the same principle, though with slightly different results, applies also to the case of a stationary source of sound and a moving observer.

The *explanation* of the Doppler effect, to use Faraday's words just quoted, we also 'think we understand pretty well'. Mechanical vibrations such as those of a siren will hit the medium air and cause it to vibrate due to its elasticity. If the source of sound issues vibrations in all directions, and if the air is at rest while the source is moving through it, then vibrations from the front of the source will hit the air at shorter intervals than those from its rear. In the converse situation, when the siren is at rest and the observer is moving, then, depending on the direction of his motion, the vibrations of the air will meet *his ear* at shorter or longer intervals. The principle is the same in both cases. Once sound has left its source, not that source but only the transporting medium, in our example air, will from then on determine its propagation and its frequencies.

Applying that principle to Fresnel's hypothesis of ether being dragged along by matter will provide a simple explanation also for the behaviour of light. Light waves emanating from the matter of some distant celestial body will hit the medium ether when leaving that source. And if that ether is being dragged along by the atmosphere of the celestial body, then there will be no significant differences in the velocities of ether relative to matter. The frequencies of light will therefore not be affected. From then on that light, consisting of oscillations of ether, will be transported through space by the medium ether, and its oscillations will be influenced only by the physical properties of that ether, whichever they are. When approaching the earth, these oscillations will continue their way through ether that is now being dragged along by the earth's atmosphere and will adapt their frequencies to the physical properties of that ether. And since that ether is being dragged along, there will be no shift of frequencie which we could observe from *within* the earth's atmosphere. Going by Fresnel's hypothesis of moving ether no significant shift in the frequencies of light is to be expected from the outset.

An even moderately unprejudiced interpretation of the evidence available at the turn of the 19th Century would thus have shown that Fizeau's and Michelson's experiments had indeed refuted the theory of stationary ether but did not clash in any way with Fresnel's hypothesis of ether being dragged along by matter. The spirit of that period would have it otherwise however. Few took interest in empirical results then, but many believed in mathematical calculations. They needed fixed operands for setting up mathematical equations because they believed physics could not survive as an 'exact science' without that. So they searched for mathematical approaches and for finding them sacrificed everything that stood in their way, including logic and empirical evidence. The following chapters will show what came thereof.

III

Before coming to that, we must take yet another look back on ether theory. History shows that its merits did not lie in tradition alone. They were due its high power of explanation which in turn was due to its clarity. Assuming the existence of an elastic medium filling all space explained many physical phenomena that would have remained inexplicable without it. They included not only the already mentioned phenomena of light such as its transportation, or its high speed, or interference and polarization. Electricity and magnetism, too, were mysteries to the understanding of which the ether hypothesis seemed at least to open approaches even if it did not provide ready-made solutions. Without assuming some transporting medium however, explaining those phenomena remained quite impossible.

The great question is why the ether hypothesis was abandoned nevertheless. The following chapters will show that the main responsibility lay

with Max Planck's quantum theory and Einstein's theory of relativity, and
with the intellectual climate which they created.

CHAPTER 8:

THE GREAT QUANTUM MUDDLE

Grave bells for ether theory; many phenomena left unexplained; confusion of descriptions and explanations. I. Stages of development of early quantum theory: Planck's starting point; atomistic interpretation of his theory; Einstein's light quantum hypothesis, Rutherford-Bohr's planetary model of the atom, quantum mechanics, Copenhagen Interpretation. II. Background of Planck's approach; conflicting formulae by Raleigh-Jeans and Wilhelm Wien describing distribution of spectral energy; Planck's 'ideal black body'; his problem of dealing with entropy solved by 'introducing probability considerations'; interpreting his approach his solution reached by interpolation; his natural constant h only for making deduction possible; experiments in his view approximations to truth. III. Einstein's photon theory shifting the meaning from 'quantum of action' to 'quantum of energy'; Planck's quantum was mathematical quantity, Einstein's physical entity; indivisibility of h never established by Planck but presupposed by Einstein without giving reason; logical effect of shifting the meaning of concepts (Thatcher-example); indivisibility of h never established by anyone. IV. The Copenhagen Interpretation of quantum mechanics; its logical problems. V. Ether theory victim to lack of criticism before and after World War I.

The grave bells for ether theory rang in the early years of the 20th Century. It fell from grace and soon found followers no longer who would muster the courage needed for standing by it.

Seen from the angle of the theory of science that development is badly in need of explanation. The phenomenon of *horror vacui* exists not only in physics but also in the theory of science, though in a figurative sense there. A theory may be as fragmentary as will be, and those working in that field may long have realised its deficiencies. As long as no one proposed a better one, providing at least the same explanations and avoiding at least some of its mistakes, they will normally stick with the old one, unsatisfactory though it may be. The long life of Ptolemaic theory shows that. Abandoning a theory without replacing it by something better would mean giving up the quest for explanations altogether. It would go against man's inborn curiosity and hence against science itself.

Ether theory was buried in spite of that. What might explain its demise? Where is the theory that will explain at least as many physical phenomena as ether theory does, and will avoid its mistakes or at least some of them? How can we explain the propagation of light without a transporting medium? How can we explain interference and polarization of light without an oscillating medium? These are but some of the questions remaining unanswered after ether theory was given up. More could be named.

Most of the misunderstandings arose from confusing descriptions and explanations. Belief in the omnipotence of mathematics had become so strong by that time that many scientists thought once a physical phenomenon had been accurately described in mathematical terms it had also been explained thereby. For showing that development in detail the chronological sequence of events is important. That is why I must begin by discussing Max Planck's quantum theory. It had no immediate bearing on ether theory, yet came to be an important prerequisite for its downfall in the early decades of the 20th Century. Only quantum theory could fill the gap which discarding the ether hypothesis would otherwise have created. It did not fill that gap with a plausible theory however, but only with the illusion of a completely determined and axiomatized science.

I

Planck's thoughts date back to the late 19th Century but his paper *On the Theory of the Energy Distribution Law of the Normal Spectrum*, which he read to the Physical Society at Berlin on December 14, 1900, is usually considered the official inauguration of quantum theory[237]. In that paper he first presented to a larger public the formula of radiation which he had developed, and introduced the 'natural constant' h standing for a 'quantum of action' in his terminology[238]. In the ensuing century quantum theory passed through several stages of development. I will describe them only in outline at first and then go into details as far as necessary.

(1) The title of Planck's paper indicates the problem he wanted to solve. He was searching for a formula describing the spectral distribution of light energy in the 'normal spectrum' which to him was the spectrum of an 'ideal black body'. Maxwell's theory presupposed that energy would increase or decrease continuously when such a body is being heated. That raised problems to be discussed further down. Purely theoretical considerations convinced Planck that the change of energy could not be continuous but must proceed in small steps, or jumps, not divisible themselves any further. His quantum theory stated no more than that at first.

Planck's theory was revolutionary on the one hand, causing radical changes in the worldview of physical science. On the other hand some interpreted it as a sequel of atomism, a philosophical tradition assuming since antiquity that the ultimate components of matter must be indivisible. It is by no means certain that Planck himself agreed with that interpretation of his theory, rather on the contrary. But in the eyes of some of his contemporaries it shifted the ancient hypothesis of classical atomism to the abstract level of mathematics without fundamentally changing the assumptions underlying it.

Understanding Planck's theory in the atomistic sense seemed natural in a way although his 'quantum of action', denominated h, neither had physical substance of its own nor other material properties. It did not stand for those last entities of which atomism assumed matter to be com-

---

237 Max Planck, *On the Theory of the Energy Distribution Law of the Normal Spectrum*, Verhandlungen der Deutschen Physikalischen Gesellschaft No. 17 (1900), p. 237 ff.(translation in *The Old Quantum Theory*, ed. by D. ter Haar, Pergamon Press, 1967, p. 82).

238 In German Planck's term was ,Wirkungsquantum'. Literally translated that would be 'quantum of effect'.

posed, but merely was an abstract quantity serving to describe material processes larger than itself. Yet it was to be a *smallest* unit of nature, indivisible itself therefore, and directly relevant for all attempts at describing anything material. That seemed to be not without precedent in atomic theory. In his theory of natural forces Roger Boscovich (1711–1787) had already described atoms on the one hand as *ideal* points but on the other hand as owning physical properties such as attractive and repulsive forces. Planck's theory also split up the whole of microphysics in smallest units. The postulate of atomism, requiring physical explanations to be reducible to smallest units not reducible themselves, thus seemed to survive in principle though under a different name.

The actual existence of those units remained in a theoretical twilight however. In Planck's theory their reality was not the result of physical properties they owned but of their mathematical necessity. He assumed his abstract 'natural constant' h to be determining the rules of all physical processes because mathematical solutions would have been impossible without that. That is why it was indivisible. Being an elementary constant of physics its 'nature' was that it could not be split in smaller units.

(2) Einstein seems to have understood quantum theory in the atomistic sense. His paper *Concerning an Heuristic Point of View Toward the Emission and Transformation of Light*[239] (1905) was the next major step in the development of the theory. The light quantum hypothesis stated there, for which he received the Nobel Prize of 1921 in 1922, was an application of his heuristic method discussed in *Chapter 4*. It meant departing from the ether hypothesis which had dominated the 19th Century. In a way it even meant going back to Newton's particle theory of light. We will see that in detail in section *III* below.

---

239   Einstein, *Über einen die Erzeugung und Umwandlung des Lichts betreffenden heuristischen Gesichtspunkt* (On a Heuristic Point of View Concerning the Production and Transformation of Light), Annalen der Physik vol. 17 (1905), p. 132 ff. I already discussed some aspects of that paper in *Chapter 4, II, 1.*

(3) The planetary model of the atom proposed by Ernest Rutherford (1871–1937) in 1911 and its integration into quantum theory proposed by Niels Bohr in 1913 then marked the third stage in the development of quantum theory[240].

Rutherford's model had emanated from the theory of electrons developed by Lorentz who had in turn been inspired by Faraday's discovery of the laws of electrolysis (1834). One assumed electrons to be connections between atoms at first, perhaps like corkscrew springs placed on the surface of the nucleus and stimulated into oscillation when induced. Their tendency to resonate was the only thing believed to be known reliably. That is why publications of that time often used the term 'resonators' where modern physicists would speak of 'electrons'[241]. Rutherford's planetary model, which assumed the atom to be repeating the order of our planetary system at the level of microphysics, was far more elegant as a visualization therefore. I will discuss it in more detail in *Chapter 10*. Bohr then integrated that model into quantum theory by postulating that the atom will emit and absorb energy through its electrons changing their orbits, and that the steps or jumps between those orbits must be calculable in integer multiples of Planck's constant h[242]. That was the origin of quantum mechanics.

(4) The so-called 'Copenhagen Interpretation of Quantum Mechanics' going back to conferences of physicists in 1927 and influenced strongly by Niels Bohr and Werner Heisenberg[243] then marks something like a break in the early development of quantum theory. In the presentation of his famous 'uncertainty relations' Heisenberg had come to the conclusion that it is mathematically impossible to determine simultaneously

---

240 The first to propose a planetary model was Nagaoka Hantarō in 1904, but his theory never connected with quantum theory in any way.

241 Apparently, Hermann v. Helmholtz and George Stoney first introduced the name 'electron'. Lorentz proposed to consider it as the carrier of electric charge.

242 Bohr, *On the Constitution of Atoms and Molecules*, Philosophical Magazine 26 (1913), p. 1-25; 476–502.

243 Bohr, *Das Quantenpostulat und die neuere Entwicklung der Atomistik*, p. 245 ff.

both position and momentum of a particle. His next inference was that 'we simply cannot know the present in principle in all its parameters' because human thought could not be sharper than nature itself[244]. Since he believed quantum mechanics to be determining all physical events and since man himself must consist of atoms and their particles, therefore, so he argued, physics will set insurmountable limits even to the *possibilities* of human knowledge. The Copenhagen Interpretation was to resolve also the problems of light raised by the fact that wave theory could explain some of its effects while the atomistic interpretation of quantum theory seemed to indicate that light must consist of particles. Niels Bohr claimed that there is a 'dualism' of physical phenomena making compatible those two aspects.

In the following sections I will deal only with epistemological aspects of the older quantum theory just mentioned. Going into later developments will be unnecessary. I have discussed some aspects elsewhere, and the arguments against the early theory refute also what came afterwards because no one ever tried again to answer the basic questions[245]. Instead, beginning with the Copenhagen Interpretation at the latest, the whole of quantum theory bogged down in what Karl Popper fittingly named the 'great quantum muddle': a methodological and epistemological chaos of truly Babylonian dimensions the unravelling of which is considered hopeless still in our days[246].

II

We must see Max Planck's original quantum theory and the underlying approaches before the background of the physical knowledge available in

---

244 Heisenberg, Physik und Philosophie (1959), p. 25.

245 For my criticism see *Popper versus Einstein*, p. 168 ff. – At that time I had not yet noticed the logical problems of quantum theory.

246 See Popper, *Quantum Theory and the Schism in Physics*, p. 4 f.

his time. The modern atomic theory was in its infancy then, and the planetary model of the atom had not yet been invented. Rutherford proposed it only in 1911.

In his search for a mathematical formula describing the spectral distribution of light energy in the 'normal spectrum' Planck therefore followed the older theory visualizing electrons as oscillating connections of atoms, absorbing energy and re-emitting it by resonance[247]. That view met with difficulties. Had it been correct, then it would have encouraged the assumption that the energy emitted by a material body must influence the frequency of its oscillations, hence its colour. At extreme heat a body should emit radiation only in ultraviolet frequencies and thus become invisible to the human eye. Conversely the frequency of light should decrease in proportion to the decrease of energy. A body losing temperature should change colour and eventually also become invisible to the human eye when the frequencies reached the infrared region of the spectrum. Such inferences were obviously incompatible with the visible effects of light. So there were problems remaining unresolved.

An experimental physicist might have approached those problems by testing in experiments the different hypotheses at stake. Planck hoped to solve them by mathematics[248]. He was convinced that immutable mathematical laws must be governing the visible and measurable phenomena of light. Therefore he endeavoured to obtain a better mathematical understanding of the spectrum of radiation. In order to prevent mere coincidence from affecting his research he could not concern himself with the spectrum of some random body or gas, but first had to establish 'ideal' conditions for his calculations. That was why he decided to investigate the spectrum of ideal black bodies.

Five years later Einstein was to take a similar approach in his special theory by presupposing the ideal conditions of uniform motions on

---

247   See for instance Planck, *Vorlesungen über die Theorie der Wärmestrahlung*, p. 100 ff.

248   For the following text I mainly rely on Max Planck, *Zur Theorie des Gesetzes der Energieverteilung im Normalspektrum*, p. 237 ff, on his *Vorlesungen über die Theorie der Wärmestrahlung*, and on Hermann, *Max Planck*, in *Die Großen Physiker'*, vol. II p. 143 ff.

straight lines, therefore of conditions nowhere to be observed in nature. Planck was following a suggestion made by Kirchhoff in his approach. At the turn of the 19ᵗʰ Century many physicists considered the spectrum of black bodies to be the ideal case of radiant energy, unaffected by specific properties of any material. Their notion was that an 'ideal' black body would reflect no rays at all but would absorb all impinging radiation. Although such bodies do not exist in nature and are anything but normal therefore, one nevertheless considered their spectrum the 'normal spectrum' for that reason. Science was interested particularly in its development when such a black body was being heated. In order to avoid random influences as far as possible, most experimenters did not use a body that really was black but the black inner walls of an enclosed cavity with only a small hole permitting to observe escaping radiation.

Figure (3) shows the results of measurements known in Planck's time. He was searching for a mathematical formula describing the distribution of energy shown there. Several formulae had already been proposed but none of them was altogether convincing. Wilhelm Wien had developed a formula yielding good agreement for high frequencies while other formulae gave better results at lower frequencies[249]. By interpolation of those equations and introducing his fundamental constant h, Planck succeeded in developing the equation now named the 'Planck formula'[250]. According to modern textbooks it is an exact mathematical representation of the curves shown in *Figure 3*[251]. Its implications need not interest us in

---

249 The so-called Rayleigh-Jeans formula had already been published then. But according to Hermann, *Max Planck*, in *Die Großen Physiker*, vol. II, p. 147 f. and Agassi, *Radiation Theory and the Quantum Revolution* (1993), p. 100, Planck did not know it. Apparently, he started from a related formula found by Kurlbaum and Rubens and later adapted his theory to the Rayleigh-Jeans formula. My further text will show that this makes no difference to the situation.

250 The formula is $u_v dv = \dfrac{8\pi h v^3}{c^3} \times \dfrac{dv}{e^{hv/k\vartheta} - 1}$.

In Planck's notation, u stands for spatial energy density, and v for the characteristic frequency of a resonator. The expression $u_n dv$ therefore designates the spatial energy density u as represented on the y-coordinate in relation to the referential quotient (marginal value) of the frequency v on the x-coordinate.

251 See for instance Gerthsen/Vogel, *Physik*, p. 545, at (11.11).

detail here. The only aspect important for epistemological discussion at this point is that Planck there introduced the constant h which he named 'quantum of action' and determined as $h = 6.548 \cdot 10^{-27} erg \cdot sec^{252}$. Due to a change of units its definition now is $h = 6.626 \cdot 10^{-34} Js$.

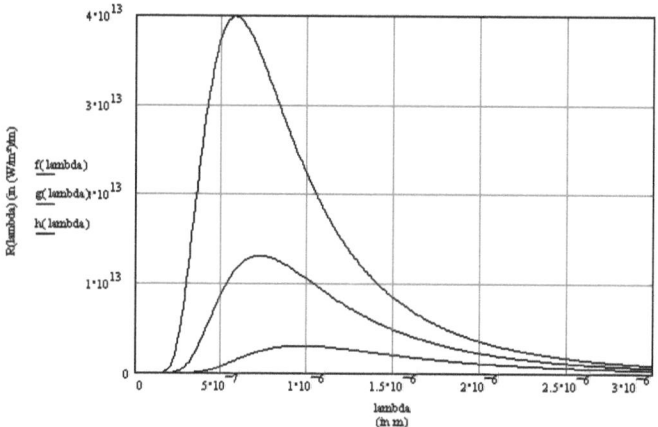

Figure 3, Spectral intensity distribution of black body radiation at different temperatures: The abscissa (x-coordinate) indicates the frequencies of radiation. The ordinate (y-coordinate) indicates the energy of radiation within the range applicable to a particular frequency. The different curves exemplify the distribution of radiation energy at the temperatures given respectively.[253]

(*1*) For illustrating the problem horizon against which we must see Planck's thoughts I now quote a short section from his already mentioned paper of 14th December 1900 *'On the Law of Distribution of Energy in the Normal Spectrum'*. Planck said there:

'Gentlemen, when some weeks ago I had the honour to draw your attention to a new formula which seemed to me to be suited to

252 Planck, *Vorlesungen über die Theorie der Wärmestrahlung*, p. 162.
253 Quoted from Christian Gerthsen/Helmut Vogel, *Physik*, 17th. ed. p. 543.

express the law of the distribution of radiation energy over the whole range of the normal spectrum, I mentioned already then that in my opinion, the usefulness of that equation was not based only on *the apparently close agreement* of the few numbers, which I could then communicate, *with the available experimental data,* but mainly on the simple structure of the formula and especially on the fact that it gave a very simple logarithmic expression for the dependence of the entropy of an irradiated monochromatic vibrating resonator on its vibrational energy. ...

Since the entropy of a resonator is thus determined by the way in which energy is distributed at one time over many resonators, I suspected one should evaluate this quantity in the electromagnetic radiation *by introducing probability considerations,* ... This suspicion has been confirmed; *I have been able to derive deductively an expression for the entropy of a monochromatically vibrating resonator and thus for the energy distribution in a stationary radiation state, that is, in the normal spectrum. ... .*

I do not wish to give today this deduction – which is based on the laws of electromagnetic radiation, thermodynamics and probability calculus – systematically in all details, but rather to explain as clearly as possible the real core of the theory. This can be done most easily by describing to you a new, completely elementary treatment through which one can evaluate – without knowing anything about a spectral formula or about any theory – the distribution of a given amount of energy over the different colours of the normal spectrum *using one constant of nature only* and after that the value of the temeperature of this energy radiation using a second constant of nature.' (My italics).[254]

The fragment describes in outline the task Planck had set himself. He had already found the radiation formula and had compared it with experimental results known at the time. That was why he spoke of 'the apparently close agreement ... with the available experimental data'. He knew the

---

254   Max Planck, *On the Theory of the Energy Distribution Law of the Normal Spectrum,* Verhandlungen der Deutschen Physikalischen Gesellschaft No. 17 (1900), p. 237 ff.(translation quoted from *The Old Quantum Theory,* ed. by D. ter Haar, Pergamon Press, 1967, p. 82).

results of experimental measurements, but his aim was to find a way of deriving his formula deductively. The passages in italics show that[255].

The problem that got in his way on that search was entropy, that is the fact that in any transformation of radiation energy into mechanical energy there will always be a small proportion that cannot be transformed. He knew the principle of entropy but not its size and could not use it as a mathematical operand therefore. That in turn would have prevented stating a formula for calculating radiation energy. Seen from the angle of mathematics, entropy was an unknown size which he could have determined only by introducing other equations.

The text quoted above indicates Planck's approach to solving that problem. 'Introducing probability considerations' was to make calculable the unknown size of entropy[256]. His constant h was the result of those considerations.

(2) Even with these first intimations Planck proved himself almost a model of the axiomatist whom we met in *Chapter 6*.

He had already found his radiation formula, and experiments had already confirmed it, but that seemed not enough to him[257]. His aim was

---

255 For that attitude see also Planck, *Zur Geschichte der Auffindung des physikalischen Wirkugsquantums*, in *Vorträge und Erinnerungen* (1947), p. 15, 24.

256 In his papers *Über das Gesetz der Energieverteilung im Normalspectrum*, Ann.d. Phys. vol. 4 (1901), p. 553 (On the Law of Distribution of Energy in the Normal Spectrum), and *Zur Geschichte der Auffindung des physikalischen Wirkugsquantums*, in *Vorträge und Erinnerungen* (1947), p. 15ff., Planck was only slightly more explicit on those probability considerations.

257 That attitude seems to have been common understanding at the time, as the following fragment from a paper by Georg Ohm shows. He began his first presentation of *Ohm's Law* like this: 'In Schweigger's Journal, I recently made public experiments that have led me to a theory of electric current, which by its *quite unsearched-for, yet perfect correspondence with experience* reveals itself as that founded in nature. Since then, I have had the good fortune of discovering *on the opposite way* from the universally acknowledged and in this field uppermost fact usually designated by the name 'electrical tension between different bodies', *with the help of mathematics*, that wonderful medium of thoughts, two laws uncovering the inner connection of all forces actively working within the galvanic chain, which seem to include within them, definitely and yet so simply, all those found previously and also what had been left open by them. Stating these factually, and showing

not applying that formula to experimental facts. That would have meant adapting its parameters which were intended for 'ideal' conditions to those of bodies existing in reality, and he wanted more than that. The formula had to be founded on deduction. Without that it seemed worthless to him because he believed that only a deductive foundation could raise it to the rank of a scientific theory. It did not bother him that his own deduction had to rely on a new constant which he introduced *ad hoc* by 'probability considerations' as an operand serving only to make the result fit. On the contrary, his faith in the necessity of deductive reasoning was so deep that he believed his constant h to have been prescribed him by nature itself. It had to be a constant of nature for the sole reason that deduction would have been impossible without it.

There was no other justification. In none of his works did he give one. Even five years later, presumably therefore after careful consideration, he presented the constant h in one of his lectures in a new equation which he introduced to the reader as a pleasant surprise in the words[258]:

'A striking feature of this result (scil. the previously established new equation) is the first appearance of a new *universal constant h* of the dimension of a *product of energy and time*.' (My translation, my italics).

In the explanations immediately following those words he wrote:

'There can be no doubt that the constant h plays a certain role in the elementary vibration processes in a centre of emission; however, for exploring this from the electrodynamic side, our previous theory *provides no further clues*. Yet, the thermodynamics of radiation will only come to a completely satisfying conclusion *if the*

---

in brief outline their application is my objective. *Showing their deduction, which will probably not be so easy, and their connection with those of related phenomena of nature, I reserve for a more detailed elaboration,* the necessary leisure for which I hope soon to find.' (Georg Simon Ohm, *Versuch einer Theorie der durch galvanische Kräfte hervorgebrachten elektroskopischen Erscheinungen*, Annalen der Physik und Chemie, vol. 82 [1826], p. 459–46; my translation, my italics).

258   Planck, *Vorlesungen über die Theorie der Wärmestrahlung*, p. 153, 154.

constant h *has been understood in its full universal significance* '. (My translation, my italics).

That was the only explanation of the universal constant h Planck ever gave. In other words *he himself knew nothing about it* because his theory 'provided no further clues'. In another paper, written in 1933 but published only posthumously in 1949, he candidly admitted his ignorance in the following words:

'The task of testing the value of the second constant, h (standig for the constant of action[259]), which was quite unconnected, appeared far more hopeless. It therefore came to me as a great surprise and delight when J. Franck and G. Hertz ... found a method ... of measuring it, which could not have been desired to be more direct. It removed the last doubt about the reality of the quantum of action.

But the most difficult problem arose now, that of giving a physical meaning to that strange constant. For, the rupture with classical theory, which introducing it caused, was far greater than I had presumed initially. ... For several years, again and again, I made attempts to fit the quantum of action into the system of classical physics somehow. *But I had no success.* In fact, we now know that shaping quantum theory was left to younger forces of whom I will only mention, in chronological order, A. Einstein, N. Bohr, M. Born, P. Jordan, W. Heidenberg, L. de Broglie, E. Schrödinger, P.A.M. Dirac.' (My translation and italics)[260].

These fragments show that Planck vaguely suspected that the constant h still needed to be 'understood in its full universal significance'. But he himself had not the slightest idea of that significance Yet he unswervingly believed that the constant h, which he had introduced *ad hoc* because

---

259 Text in brackets added.

260 Planck, *Zur Geschichte der Auffindung des physikalischen Wirkugsquantums*, in *Vorträge und Erinnerungen* (1947), p. 15, 27.

then-known measurements could not be expressed in a mathematical formula without it, must be a 'universal constant' for that very reason..

The fact that his radiation formula had already been confirmed empirically appeared to him irrelevant in that context. He mentioned it only
in passing in the introduction of his paper. Further down he even said[261]:

> "It would of course be very *complicated* actually to *carry out the
> calculations* specified although it would *certainly not be without
> interest* to examine in a simple case the degree of *approximation
> to truth* that can thus be achieved." (My translation, my italics.)

He could hardly have expressed more clearly that experiments were to him
of secondary importance at best. They were 'not without interest' but no
more than that. They certainly did not give rise to making 'complicated
… calculations'.

My italics in the text last quoted even indicate that Planck believed
'approximations to truth' to lie not in theories but in experiments. Theories, as I understand him at this point, did not approximate truth. They
were truth. They alone were legitimate science to him. Everything else
was mere 'opinion'. His axiomatic understanding presupposed that in an
'exact science' like physics the theory had been inferred by the strict rules
of mathematics from premises that were indubitable. Experiments might
help to find true theories in this performance. But their results were mere
indicators in his view, assisting the weak human mind in its search for correct deduction. An experiment however could approximate truth only if
it *confirmed* the theory that had been found deductively. He presupposed
that it must and would confirm theories that had been deduced correctly,
believing that to be self-evident. That was also why introducing a mathematical constant that united two otherwise contradicting formulae in one
comprehensive formula appeared to him a magnificent discovery of science. And also for that reason, experimental confirmations were to him at
best 'not without interest'. On no account, so he must have thought, could

---

261 Planck, *Vorlesungen über die Theorie der Wärmestrahlung*, p. 242.

the outcome of an experiment call into question or even refute a theory that had been justified deductively. The idea that it might annihilate a theory as completely as Rœmer's discovery of the finite speed of light had annihilated Descartes' theory of its instantaneous propagation must have been completely alien to him.

(3) From today's perspective it hardly is possible anymore to imagine the world of thought in which Planck lived. It might become easier if we consider what he regarded as physical experiments.

Planck was one of the first academic teachers of physics to work only theoretically from the beginning of his academic career. He probably lacked the practical view to some extent and could never free himself from the fascination of geometry, not even when considering physical experiments. The text put in italics in the passage last quoted indicates that to him experiments seem to have consisted in 'carrying out calculations'. They were like drawing geometrical figures on paper according to the rules of geometry. The pencil must be thin, the ruler straight, and the compass must have sharp points. Given such conditions, careful handling of the instruments would yield good approximations to geometric figures without quite reaching their ideal. That would also explain why Max Planck thought it important to carry out experiments under 'ideal' conditions. An experiment could be meaningful only if it met such conditions. And if it did, then it was bound to confirm the theory that had been found by deduction. Primacy clearly belonged to theory in his view. Experiments were but secondary confirmations.

Einstein himself should have firmly contradicted such notions. I remind once more of his words[262]:

> "Insofar as the propositions of mathematics refer to reality, they are not certain, and insofar as they are certain, they do not refer to reality."

262 Einstein, *Geometry and Experience.* I quoted the text extensively in *Chapter 1, III,1.*

Since Karl Popper's *Logik der Forschung* (1934) at the latest we know that the epistemological approach underlying Planck's views is untenable. We saw that in *Chapter, 2, II*. Heisenberg, who repeated Popper's theory of science without mentioning its author, should also have disagreed with that approach therefore[263]. Yet he remained an unflinching champion of quantum theory up to the last. The actual situation seems to have been that neither he nor Einstein ever noticed the inconsistencies in Planck's thoughts. After the considerations in the *First Part* readers can now figure out themselves the reasons for that. Much of what I said there about Einstein applies also to Heisenberg in principle.

III

Only five years after Max Planck had officially launched quantum theory on December 14, 1900, Einstein took the next major step in its development. In his paper *Concerning an Heuristic Point of View Toward the Emission and Transformation of Light* (1905) he stated his light quantum hypothesis for which he later received the Nobel Prize for 1921[264]. Other than his theory of relativity it gained acceptance only with difficulty. Resistance seems to have come not least from Planck himself[265].

---

263 Heisenberg's paper *Die Bewertung der ‚modernen theoretischen Physik* (Zeitschrift für die gesamte Naturwissenschaft, 9, p. 201–212; also in *Deutsche und jüdische Physik*, p. 90 ff.) was written in 1940 and published in 1943. In that paper, he repeated in content the most important thoughts of Popper's *Logik der Forschung* (1934), in particular those on the hypothetical character of scientific theories, on the asymmetry of verifiability and falsifiability, and on the connection between objectivity and experimental testability. – Mentioning Popper as the author might have seemed inopportune under the reign of National Socialism but in later publications Heisenberg could have corrected the impression he had created, for instance in *Der Teil und das Ganze, Gespräche im Umkreis der Atomphysik* (first published in 1973).

264 Annalen der Physik 1905, p. 132 ff; the quotation following in the main text is on p. 133.

265 See Fölsing, *Albert Einstein - Eine Biographie*, p. 170; Hermann, *Max Planck* in *Die Großen Physiker*, vol. II p. 149.

On closer consideration that was hardly surprising because Einstein had radically changed the meaning of quantum theory by the way in which he made use of it in his own new theory. Planck never addressed that change directly to my knowledge, but he must have felt it nevertheless and it must have worried him at heart. Einstein's light quantum hypothesis of 1905 was in fact the first profound revolution in the history of quantum theory even if that remained unnoticed at the time. No theoretical physicist seems to have realised that with that theory Einstein had given up again the premises on which Max Planck had based his theory in 1900.

(*1*) We saw above that Max Planck had conceived the constant h as a 'product of energy and time'[266]. It had no physical properties of its own but was an abstract mathematical operand serving for describing physical processes. He himself had coined the term 'quantum of action'[267]. The name was important to him because it indicated that the constant h was independent of specific applications. That was why he wrote[268]:

> 'I want to denominate this (*scil.* the natural constant h) as the "elementary quantum of action" or as the "element of action" because it is of the same dimension as the quantity to which the principle of smallest effect owes its name.' (My translation).

The abstract quality of Planck's constant has implications which we must consider in detail.

In our time almost every physicist will take for granted that the quantum is indivisible. Even with non-physicists that is almost common knowledge; to many it is all they believe to know of quantum theory. They probably assume that its indivisibility had been carefully 'justified' by Planck himself. For it seems a matter of course that a discovery as fun-

---

266 See the quotation from Planck, *Vorlesungen über die Theorie der Wärmestrahlung*, p. 153, 154, in Section *II, 2*.

267 As mentioned above, Planck's German term was 'Wirkungsquantum'. Literally translated, that would be 'quantum of effect'.

268 Max Planck, *Vorlesungen über die Theorie der Wärmestrahlung*, p. 154.

damental as that of a last indivisible entity of nature must have been based on thorough investigation. How else could one rely on the existence of so fundamental an entity?

In spite of that however, no confirmation is to be found for that assumption in Planck's writings, not even the slightest! The allegation may sound incredible but I can only request those doubting it to check its accuracy by reading Planck's texts. There is no other way of proving it because there is nothing in those texts that I could quote for doing so. Yet the fact that he never gave reasons for the indivisibility of the constant h is of utmost importance for assessing the whole devlopment of theoretical physics in the 20th Century.

Planck's paper *On the Law of Distribution of Energy in the Normal Spectrum,* published in 1901, is slightly more elaborate than the paper he read to the Physical Society at Berlin on December 14, 1900. But even in that later paper he simply introduced h together with his other constant k by the words:

' ... and consequently ... here h and k are universal constants.'[269]

He gave no other reason. Closer consideration shows that even to search for a justification of the indivisibility of h would have been almost directly against his beliefs. The only sentence coming anywhere near to *arguing* for the indivisibility of h that I found in his writings reads like this[270].

'The fact that the constant h is introduced as a *certain finite quantity* is characteristic of the whole theory here developed. If one would assume h infinitely small, one would come to a law of radiation resulting as a *special case* from the general one (cp. the Rayleigh law ...).' (My translation; my italics).

---

269 My translation from Max Planck, *On the Law of Distribution of Energy in the Normal Spectrum,* Annalen der Physik vol.4 (1901), p. 561 (§ 10).

270 *Vorlesungen über die Theorie der Wärmestrahlung,* p. 156.

This fragment shows that in Planck's understanding the abstract aspect came first. The constant h had to be a 'certain finite quantity' because only in that case could his equations be expressing a universal law whereas assuming h to be infinitely small would have implied a continuous increase or decrease, and the equations would then have applied only to the 'special case' of the Rayleigh law. That was why he introduced the quantum of action as a 'constant of nature' and repeatedly added the epithet 'elementary'[271]. Nature itself must have prescribed it, so he must have believed, because it was mathematically indispensable for reconciling with each other the conflicting theories of Wien's equation and the Rayleigh-Jeans formula. Had h been variable, then his formula would have described only a 'special case'. And because only a constant would express a universal law of nature he believed it to be elementary. The circularity of that reasoning did not worry him. Being an adherent of the axiomatic method he believed starting from fixed basic truths to be inevitable. Without them, so he must have thought, there could be no 'exact science'. And if one had to start from them, then they must be part of nature itself, therefore existing in reality.

From Planck's point of view, as I understand him, the problem was about a methodological principle. In his notion the constant h had to be indivisible because it had to be a constant. The question of whether it was indivisible *empirically* never presented itself. It could not present itself because he did not consider the constant h an empirical entity but an abstract quantity resulting from mathematical necessity. Its elementary quality was no result of observable physical effects but of the fact that deducing the radiation formula would be impossible without it. If he had been searching for a justification of its indivisibility, then he would have been contradicting his own beliefs. He would have expressed doubts he did not harbour.

Least of all would Max Planck have resorted to experiments. They would have presupposed physical effects that could be tested, and he avoided allegations of that kind. Theory was his domain, and experiments

271 *Vorlesungen über die Theorie der Wärmestrahlung*, p. 154, 156, 162,

seemed secondary compared to that. The best they could achieve from his point of view were approximations to truth. But his natural constant h was to be a universal constant for purely mathematical reasons and it had to be applicable to all other physical quantities. That is why he never used the term 'quantum of energy' his writings but only that of a 'quantum of action'. And never in any of his writings did he speak of '*quanta* of action' because being an abstract, a mathematical constant could not exist in several specimens.

(2) Einstein did not give up Max Planck's approach explicitly in his paper on the light quantum hypothesis but he set aside its meaning from the outset. At the very beginning of that paper he proposed to treat the quantum hypothetically as something existing in reality and then to consider how far that hypothesis would carry. That was his application of the heuristic method which we discussed in *Chapter 4* in the context of his epistemological beliefs. At the same time, unfortunately, it was also where the fundamental shift in the conceptual content of quantum theory came in which messed it all up again.

The following brief sections quoted from the introduction to Einstein's to paper are symptomatic. He wrote there[272]:

> 'According to the Maxwellian theory, energy is to be considered a continuous spatial function in the case of all purely electromagnetic phenomena including light, while the energy of a ponderable object should, according to the present conceptions of physicists, be represented as a sum carried over the atoms and electrons. The energy of a ponderable body cannot be subdivided into arbitrarily

---

272 Einstein, *Über einen die Erzeugung und Umwandlung des Lichts betreffenden heuristischen Gesichtspunkt,* p. 132. - My quotation is from the official translation in American Journal of Physics, v. 33, n. 5, May 1965 (Concerning an Heuristic Point of View Toward the Emission and Transformation of Light), but at the end of the first paragraph I have added the word '(over)' which seems to have been omitted. At the end of that paragraph, Einstein's German text has the word 'Summe,' which should correctly be translated by 'sum'. Translating it by 'volume' in the official translation is better than the German original because it does not show that Einstein was not keeping apart mathematics and physics.

many or arbitrarily small parts, while the energy of a beam of light from a point source (according to the Maxwellian theory of light or, more generally, according to any wave theory) is continuously spread (over) an ever increasing volume.

...

It seems to me that the observations associated with blackbody radiation, fluorescence, the production of cathode rays by ultra-violet light, and other related phenomena connected with the emission or transformation of light are more readily understood if one *assumes* that the energy of light is discontinuously distributed in space. In accordance with *the assumption to be considered here*, the energy of a light ray spreading out from a point source is not continuously distributed over an increasing space but consists of a finite number of *energy quanta which are localized at points in space, which move without dividing, and which can only be produced and absorbed as complete units.* (My italics).

This passage has two aspects which must interest us here, one method-ological, the other conceptual. Keeping them apart will be important in our context.

(*a*) First about method. Einstein said he was proposing an 'assumption'. He assumed 'that the energy of light is discontinuously distributed in space'. He did not start from a truth believed to be certain and indisput-able therefore but on the contrary from a mere hypothesis. Its justification was to be examined only afterwards by its results.

Planck and Einstein were writing at almost at the same time as we saw, but from the point of view of methodology the difference between them was remarkable. Planck had already found his radiation formula but considered it unsatisfactory as long as it stood on empirical findings alone. That was why he searched for a way of deducing of it *from other ax-ioms* and assumed the constant *h* to be one of them. Einstein by contrast refrained from basing his hypothesis on any reasons at all. He relied only on the power of persuasion of the deductions that would *follow* from it. His heuristic method thus seems to mark him as a modern proponent of critically rational science in the sense of Popper's theory, a scientist taking up new hypotheses unprejudiced and criticizing them only subsequently at the hands of experience.

(b) That remarkable progress however was forthwith undone again by the inconsistency of Einstein's thoughts. He converted Planck's abstract 'quantum of action' into a plurality of 'quanta of energy'. In his understanding they stood for amounts of energy 'localized at points of space' as the text just quoted shows, and they could even 'move without dividing'. They were equipped with properties belonging only to things that are part of physical reality. Thus, while Planck's constant h stood for an abstract quantity, Einstein's quantum of energy clearly stood for an empirical entity[273].

Compared to Planck's theory Einstein had significantly shifted the meaning of the term 'quantum' in his essay. It was a kind of mistake that repeatedly happened to him in his writings. That is why I must briefly demonstrate by an example the devastating consequences arising from such operations, ostensively logical but in fact consisting in a shift of meaning in of one of the terms used in the operation without changing the term itself. The sentences

(P_1) All ladies have brains
(P_2) Margaret Thatcher was an iron lady
(C_1) Margaret Thatcher had brains

show the deduction of a true conclusion (C_1) from true premises (P_1, P_2). The logical operation underlying them follows closely the syllogism 'Barbara' of classical logic. It is formally correct. The truth of the premises implies that of the conclusion drawn from them.

---

273 This becomes apparent also at the end of his paper where Einstein spoke of 'light quanta'. He began his discussion of Planck's theory by speaking of 'Elementarquanta' (p. 136), then introduced 'movable points of any kind' (p. 141), then proposed that light might consist of 'energy quanta' that are independent of one another (p. 143), and ended up by speaking of 'light quanta' that could be absorbed or cause ionisation of molecules (p. 148).

The formalism of deduction remains the same in all respects if we write:

(P$_3$) All iron bodies are electromagnetic.
(P$_4$) Margaret Thatcher was an iron lady.
(C$_2$) Margaret Thatcher was electromagnetic.

That sequence also follows the syllogism 'Barbara' of classical logic. Its formal aspects differ in no relevant way from the first. The truth of the second premise (P$_4$) also is out of question. Margaret Thatcher was well known as the 'Iron Lady' in her time.

Yet the second conclusion (C$_2$) is false so obviously that no formal logic could conceal that. The fault is not in the logical formalism however. In that respect there is no difference to the first sequence. Nor was the fault caused by one of the premises being wrong. Both premises (P$_3$, P$_4$) are true also in the second sequence.

The second deduction is fallacious because the *meaning* of one of the terms used in the two premises (P$_3$, P$_4$) *shifted without the term itself being changed*. In the first premise of the second case (P$_3$) the word 'iron' refers to a chemical element whereas in the second (P$_4$) it stands for certain human characteristics of the late British Prime Minister.

That shift of meaning was not introduced explicitly. It came without special mention all by itself from the conceptual background of the premise P$_4$ which presupposed as being common knowledge that Margaret Thatcher was nicknamed the 'Iron Lady'. It was at the level of mental associations and it shows that logical operations may lead to false conclusions even if they were based on true premises *and* consist in a formally correct application of the rules of logical inference.

Logical mistakes happen not only at the level of abstract concepts or rules of logic. In fact, they rarely happen at that level, especially after the invention of computers that make no mistakes at the level of the binary code. In real life logical mistakes mostly arise from situations where some conceptual content had been allocated to some term or symbol used in a statement and the allocation was not upheld in the further course.

The process of connecting some concept with a specific meaning thus creates a source of error to which scientists must pay careful attention, especially when expressing their ideas in mathematical formulae. For, as we saw in *Chapter 5, II,* the application of the rules of logic consists *only* in allocating to the words or characters occurring in language or in mathematics some specific meaning which we want them to convey in some context, and in *retaining* that meaning in that context. Words such as 'iron', or abstracts such as *a, b, c* or *x, y, z,* will thus be connected with sets of facts, for example by denoting the velocity of light by *c,* or energy by *e,* or Planck's constant by *h.* That is what gives them their meaning. They stand as symbols for something not identical with themselves. With every application of that kind we leave the field of 'pure' mathematics or of 'formal' logic. And by doing so we create a source of error that did not exist there before.

(*c*) Einstein repeatedly committed mistakes of that type. Even in the case of quantum theory he used identical expressions in different meanings within one context, placing himself thus on the level of my Thatcher example. He shifted the meaning of his concepts without declaring that and took back half of that shift the next moment, making things even worse thereby.

In his light quantum theory he converted Planck's singular abstract 'quantum of action' to a plurality of 'energy quanta which are localized at points in space' and which could even 'move without dividing', thereby transforming it from an abstract quantity to an empirical entity. Yet he retained *part* of Planck's original meaning by continuing to assume that his newly introduced 'energy quanta' could 'only be ... absorbed as complete units'. Thus Einstein *in fact introduced the indivisibility of his 'quanta of energy' without stating any reason for that, and even though no one else before him had ever given any reason for their being indivisible.*

Planck never discussed indivisibility as we saw. In his view the quantum of *action,* standing for a product of energy and time[274], was an ele-

---

274 Planck, *Vorlesungen über die Theorie der Wärmestrahlung,* p. 153, 154.

mentary constant of nature, its indivisibility needing no further justification therefore. Einstein seems not to have noticed that however. At any rate he did not introduce his assertion that the quanta of energy could 'only be ... absorbed as complete units' as a hypothesis to be tested by experiment. Instead, he gave the impression as if they still were the same entities as the one introduced by Planck. He must have believed that Planck had justified the indivisiblity of his quantum of action deductively.

That belief would have been mistaken. If Planck had defined his natural constant h as standing for a quantity of real energy, therefore as an empirical entity, then he would have had to explain from the outset why he assumed it to be indivisible. Why after all should there be a limit to the smallness of energy impulses? Had not Maxwell's theory started from exactly the opposite assumption that energy will increase or decrease continuously and can be divided infinitely? Yet even if there were a limit to divisibility, why should it be at $h = 6.548 \cdot 10^{-27} erg \cdot sec$? Why should we, mere human beings with but poor sensory faculties, be able to recognize that limit *empirically*? Why should it apply to other things besides the heating of ideal black bodies? All that could hardly be explained to anyone at first glance. Considerable efforts would at least have been needed for making it credible to physicists or other readers. And the probability of convincing a majority of scientists would have been extremely small.

It never came to that however. Due to his axiomatic understanding of science Planck never thought it necessary to state any reasons at all for the indivisibility of h. And Einstein, in the introduction of his paper, summarily and without any explanation presupposed the quantum to be indivisible and not to be *designating* an abstract quantity of energy but *being* energy itself. He confused language and reality or mathematics and physics at this point, and thereby performed the uncontrolled shift of meaning demonstrated in my Thatcher example, invalidating thus all other conclusions drawn from his hypothesis of the existence of a quantum of energy.

In Einstein's theory Planck's *calculatory* entity of the fundamental constant h became the *physical* entity of a concentration of energy which he assumed to be the smallest possible in nature. Without any justification he denaturized Planck's 'quantum of action' into a plurality of 'quanta of energy' and that was not all. Einstein's energy quanta even could move about in space as the text last quoted shows. They were nothing but light particles. At the end of his paper he himself even spoke of 'light quanta'[275], indicating thus that they were the 'photons' of modern terminology. At bottom his light quantum hypothesis thus meant returning to Newton's particle theory of light which had been considered refuted in the 19th Century because several physical phenomena of light could be explained satisfactorily only through a wave theory relying on the ether hypothesis.

(d) From the point of view of scientific method there would have been no objection to resuming the old particle theory of light provided Einstein had done it openly and explicitly. Qualified discussions of conflicting theories of physics are the very motor of science. Interesting new ideas often emerged in their course. An open dispute between adherents of the wave theory of light relying on the ether hypothesis and those proposing a new variety of Newton's particle theory of light might have been very stimulating[276].

Einstein's denaturation of quantum theory was neither explicit nor open however. The shift of meaning from which he started in his new theory was well hidden in premises which he assumed to be true without explaining them. He probably never noticed that with his light quantum hypothesis he already had left again the ground of Planck's theory. In his paper *Concerning an Heuristic Point of View Toward the Emission and*

---

275 pp. 144–147

276 The problem itself is interesting, and the *second* sentence of the text quoted above in section *III, 2* from Einstein's paper *Über einen die Erzeugung und Umwandlung des Lichts betreffenden heuristischen Gesichtspunkt*, p. 132 (Concerning an Heuristic Point of View Toward the Emission and Transformation of Light) indicates that he had seen it. If the energy of light *could* be subdivided infinitely while the energy of a ponderable body could *not* be subdivided infinitely, then a *mathematical* representation of those two theories would lead to a contradiction. The new variety of the ether hypothesis, which I will propose in *Chapter 11*, will solve that problem by explaining energy through the elasticity of ether, and by permitting longitudinal as well as transversal waves. I will mention it there in a footnote to section *V*.

Transformation of Light (1905) he did not mention Planck's natural constant h even once. Accordingly, the size of the quantum of action which Planck had calculated at $h = 6.548 \cdot 10^{-27} erg \cdot sec$ appeared nowhere in that paper. That could hardly have escaped Planck's notice but he seems to have been unable to articulate his qualms clearly and to communicate them to others.

We thus face the astounding fact that from Max Planck's first presentation of the quantum hypothesis in 1900 to the present day *no one ever established the indivisibility of the quantum*. Planck himself never established it because he was firmly convinced that his quantum of *action* was an elementary quantity based on mathematical deduction. Einstein however believed justifying indivisibility to be unnecessary because he relied on Planck's deduction which did not exist. Nevertheless he shifted the meaning of the concept 'quantum' by transforming Planck's *abstract* quantity into a plurality of *physical* entities which he then named 'quanta of energy' and treated as such.

Experimental investigation would have been of utmost importance in that situation. Yet *no one ever tested the indivisibility of the quantum by experiment either*[277]. In fact, I know of no experiment by which we could put it to the test even in our time. Einstein himself did not notice the shift of conceptual meaning which he had caused, and others followed him uncritically, among them the Nobel committee in 1922. The hypothesis of the quantum being indivisible thus evaded discussion for more than a century.

The change of meaning implied an exchange of the epistemological approach while at the same time changing the key concept of the theory. That was why Einstein's paper on the light quantum hypothesis did not keep in the end what he promised at the beginning. He did not follow

---

277 Experiments for testing quantum theory were carried out among others by Aspect et al. (Aspect, Grangier, Roger, *Experimental Realization of Einstein-Podolsky-Rosen Gedankenexperiment: A new Violation of Bell's Inequalities*, Phys.Rev.Lett. 49 [1982], p. 91; Aspect, Dalibard, Roger, *Experimental Tzest of Bell's Inequalities using Time-Varying Analyzers*, Phys.Rev.Lett. 49 [1982], p.1804). They did not aim at testing the indivisibility of the energy quantum however. For details, see my *Popper versus Einstein*, p. 172ff.

to the end the correct methodological approach taken by his heuristic method. We already saw that in *Chapter 5, III.* He began by introducing a new hypothesis, promising it would make better understood some physical phenomena, but he then went on from there by pursuing the purely mathematical approach seemingly following from the axiomatic method. That was why in his further reflections he lost himself in mathematical calculations showing no relation to empirics any more.

The shift of meaning of quantum theory remained unnoticed throughout the 20th Century. It is at the bottom of the methodological chaos which Karl Popper later called 'the great quantum muddle', a confusion which nobody has been able to unravel again ever since[278]. The fact that no reason is known why we should believe the quantum to be indivisible ought to put an end to all that.

IV

The question of the indivisibility of the quantum is not the only problem raised by quantum theory. The concept of the 'quantum' itself raises an even greater one which best becomes visible in the theory of quantum mechanics mentioned above in section *I, 3*. We must turn to the so-called Copenhagen Interpretation of Quantum Mechanics for seeing that.

As mentioned, Niels Bohr had tried to bring in line quantum theory and the planetary model of the atom model by postulating that the atom will emit and absorb energy through its electrons changing their orbits, and that the steps between those orbits must be calculable in integer multiples of Planck's constant h[279]. Quantum mechanics is about the inferences following from that postulate.

---

278 Popper, *Quantum Theory and the Schism in Physics* (1982) p. 4 f.

279 Bohr, *On the Constitution of Atoms and Molecules*, Philosophical Magazine 26 (1913), p. 1–25; 476–502.

The Copenhagen Interpretation takes its name from a conference of physicists at Copenhagen in 1927 dealing mainly with papers published earlier by Niels Bohr, Werner Heisenberg, Max Born and Pasqual Jordan. Other than Einstein had done in his light quantum hypothesis, those physicists did not assume the quantum itself to consist of energy, yet they believed it to be relevant for any change occurring in matter. In the Copenhagen Interpretation the quantum came to be an abstract quantity again. In a way that meant going back again behind Einstein's light quantum hypothesis to Planck's original theory. Yet the meaning of the term 'quantum' in the Copenhagen Interpretation was still not the same as that in Planck's theory. It did not stand for a unit describing *energy* but was assumed to be relevant somehow for calculating the *distance* of the quantum jumps. The Copenhagen Interpretation was in fact the third different interpretation of the meaning of the term 'quantum'.

Planck's theory assumed changes of energy in nature to happen in steps which take place *instantaneously* and can be described in integers of the constant *h*. Bohr spoke of an 'essential discontinuity' in that context. Heisenberg also cearly recognized the problem arising from the instantaneousness of the quantum jump[280]. It does not arise in a theory assuming the change of energy to be a continuous process. Maxwellian theory can describe change mathematically as a physical process which we can visualize in curve charts. Planck's quantum jump however is no physical process in that sense because it is instantaneous. It does not take *time*.

The Copenhagen Interpretation was to solve the problems arising from that. Its participants tried to tackle them at the level of philosophy or at that of mathematics. Bohr assumed a 'dualism' of waves and particles which he named 'principle of complementarity', and Heisenberg developed his famous uncertainty relations for solving it.

Observing the principles of methodological nominalism would have been better. We can use physical parameters (among other things) for

---

280 Bohr, *The Quantum Postulate and the Recent Development of Atomic Theory* (in: *Atomic Theory and the Description of Nature*), p. 53; Heisenberg, *Physik und Philosophie* (1959), p. 25.

describing events, or states, or processes. In physical understanding the definition of the word 'process' implies that a complete description of a process must include the parameter of time whereas the definition of the word 'state' implies that time is no relevant parameter for describing the state of something. A physical body is a state in that sense, at least if we consider it only by itself and assume it to preserve its identity for a while. Being a state, a physical body cannot be in two places at once by physical definition because identifying it would then be impossible. Some may believe in the possibility of a body being in two places at once but in that case they would either be using the word 'body' in a different meaning than the one just given or they would be discussing things not at the level of physics but at that of metaphysics.

If we use those physical definitions also when describing the quantum jump of particles, then we must make up our minds whether we want to describe those particles as physical bodies or as processes. If we describe them as physical bodies, then they cannot be in two places at once, and if we describe them as processes, then that alone would not tell us what is going on. We would still have to find out what is jumping and how it can do so without taking time.

Born, Jordan and Heisenberg however thought they could overcome the difficulties raised by the intantaneousness of the quantum jump by way of mathematics. Planck had introduced 'probability considerations' for deducing his constant h. Their approach came quite close to that. For reconciling with quantum mechanics the self-contradiction underlying Planck's theory, consisting in assuming a change of location without assuming motion, they introduced the calculus of matrix algebra which permits to include in one matrix various states of its members. By that stratagem they in fact re-introduced in their mathematical concepts the parameter of time which they had eliminated from the situation they were trying to describe. That clearly shows at the beginning of one of the early papers by Born and Jordan where they wrote[281]:

---

281   Bohr, Jordan, *Zur Quantenmechanik*, Zs.f.Phys. 34 (1925), p. 858.

'The only restriction which we impose upon the choice of coordinates is to base our considerations upon libration coordinates, *which in classical theory are periodic functions of time.*'[282] (My italics).

That sentence shows that they used coordinates describing 'functions of time' for describing quantum jumps in which time was to be no relevant factor. The effect of that stratagem was that the difficulty raised by the instantaneousness of the quantum jump was no longer visible in the description of that jump but was hidden in the uncertainty of the libration coordinates used for making it calculable[283]. The *physical* problem however, consisting in the question how a particle can move from one place to another *without its motion taking time*, remained unresolved. They simply evaded it. Like Einstein they must have firmly believed that to all problems of physics there must be solutions at the level of mathematics. In fact however, they were putting the cart before the horse.

V

Compared to those mental excesses Einstein's light quantum hypothesis at least had the merit of being a theory of physics. It was an attempt at solving a physical problem by telling us something new about *reality*. That is why it should have given rise to critical scrutiny but that never came.

In Germany after the First World War (1914–1918) criticising Einstein was undesirable. In the general moral and economic collapse of the country the hopes of the German government were pinned on him. His

---

282   Translation from *Sources of Quantum Mechanics* by van der Waerden (1967), p.277.

283   A later paper by Born, Jordan, Heisenberg (*Quantenmechanik II*, Zs.f. Phys. 35 (1926) 557–615) expressed the same by saying: 'It was found possible to extend the above theory to systems having several degrees of freedom (Chapter 2), and by the introduction of "canonical transformations" to reduce the problem of integrating the equations of motion to a known mathematical formulation.' (Translation from *Sources of Quantum Mechanics* by van der Waerden (1967), p.321).

achievements were to bring back to German science its international reputation. When the British Royal Society officially accepted his general theory of relativity after Eddington's expedition of 1919, that came as a godsend not only to Einstein but also to Germany. Criticism was hardly to be expected then from German scientists struggling for subsistence. But the logical inconsistencies of the light quantum hypothesis remained unnoticed also abroad. And when Einstein had been awarded the Nobel Prize for Physics of 1921 for just that theory, and had been invited to lecture on it at Princeton in 1932, its truth was officially confirmed as it were. Any criticism that might still have been forthcoming was nipped in the bud.

It goes without saying that this development could not be beneficial to ether theory. It simply stood no chance any more. Einstein's theory of light agreed with the views held by scientists believing in axiomatics because it seemed to endorse their claim to scientific exactness. And Einstein promised to solve the problems of light without falling back on ether theory. Whosoever had not noticed his deviation from Planck's quantum theory – and it seems that no one noticed it at the time – was led to believe that he had inferred his theory of light by cogent mathematical deduction from unquestionable premises which meant that it could lay claim to the certainty of logic. It seemed to open up the possibility of reducing all physical phenomena to smallest, well-defined units of $h = 6.548 \cdot 10^{-27} erg \cdot sec$, and of thus restoring the ancient unity of mathematics and physics for which so many physicists were yearning. Those following it were 'modern' therefore. When even the Nobel Committee had given its blessing most scientists would probably have doubted themselves more readily than Einstein's theory.

Anyone standing openly by ether theory in that situation would have stigmatized himself as being hopelessly out of fashion. Max Planck's quantum theory and its denaturation by Einstein's light quantum hypothesis together had prepared an intellectual climate in which there simply was no place for ether theory any longer. The great success of the theory of relativity which we will discuss in the following chapters then finally spelled its doom.

CHAPTER 9:

ETHER THEORY VERSUS SPECIAL RELATIVITY

'An Iron Curtain has descended.'

SIR WINSTON CHURCHILL (1874–1965)

The neutrino experiment of 2011; measurements ended when showing subluminal velocities; STR is the Iron Curtain of Science; culmination point of theoretical physics with conflicting goals. I. Einstein's ambiguity on ether theory; its problems of credibility; Maxwell's toleration of ether theory; Michelson's experiments misinterpreted as attempts at verifying ether; Popper's theory still unknown; Einstein's pincer attack on theories of light and moving bodies; decline of ether theory due to attitude of uncritical rationalism. II. Contradictions in STR; 'proper time' of moving satellite self-contradicting; also if using other standards; Planck and Einstein using unknown as a benchmark for measuring the known. III. Methodological aspects of STR; Einstein's heuristic approach; omissions in his paper: all counter arguments left unmentioned, Young's double slit experiment, interference, polarization, Huygens' and Faraday's analogy to sound, problem of transportation without carrier; Einstein not interested in explanations but only in mathematical approaches. IV. Einstein's mathematical self-contradiction; all results of STR arbitrary; history of its discovery; Einstein's premise of constancy refers only to emitting body; insofar agreement with sound analogy; his equations assume constancy also relative to receiving body; his definition of synchronism explicitly applied also to moving system; contradiction refutes entire STR; mathematical mistake located in equation (2) defining speed of light in moving system; same mistake in other derivations of the Lorentz transformation. V. On the reality of time; applying methodological nominalism. VI. On the Relevance of e = mc²: reputed to have established equivalence of mass and energy and explained nuclear fission; based on same self-contradiction as STR, therefore all results random; Einstein also misunderstood energy; his deduction circular because presupposing inherent energy of bodies; his mistake originates in Maxwell's theory; probably because

Newton did not mention Huygens' theory of energy transport; interchangeability of matter and energy is real problem; laws of conservation are not laws of nature but instruments needed for applying mathematics.

In September/November 2011 the CERN[284] reported of superluminal velocities of neutrinos having been measured on a distance of $730km$ between Geneva and the Gran Sasso in Italy. The report caused deep surprise in the world of theoretical physics because Einstein's special theory of relativity declares superluminal velocities to be impossible. In its press release the CERN cautiously referred to them as an 'anomaly'[285]. Measurements were repeated several times and with various different settings. At first they again showed superluminal velocities and were continued. Later experiments then showed velocities equal to or lower than that of light, and the initial measurements were 'attributed to a faulty element of the experiment's fibre optic timing system'. What at first seemed a serious challenge to the special theory on empirical grounds thus ostensively resulted in triumphantly confirming it.

I sometimes wonder what the attitude of the world would have been if the scientists at the CERN had made those measurements in times of an impending war or while the arms race between the United States and the Soviet Union still was at large. In 1938/39, when Hitler was in power and Hahn and Strassmann had shown nuclear fission to be feasible, the U.S. Government were not slow to recognize the military relevance of that discovery. Would their attitude have been different if some scientist had discovered that there are waves or particles so fast that they can overtake even light signals? Would the generals at the Pentagon in the times of Leonid Breshnew also have trusted in superluminal velocities being impossible just because Einstein had said so? Or would they have preferred to be on the safe side and let measuring go on until it was better known whether truth was in theory or in experiments?

---

284 'Conseil Européen pour la Recherche Nucléaire'.

285 CERN press release of September 23, with updates of November 18, 2011, and February 23, 2012.

I am not advertising military armament by this, but just intimating that in a science unaffected by the dogma of special relativity the initial results of those neutrino experiments would not have been considered an 'anomaly' but an exciting *discovery*. They would have indicated that there might still be phenomena unknown to physical science. The whole world of physicists would have been curious therefore, and burning to see the results of those experiments *confirmed*. Under such conditions they might even have marked the beginning of a new era of science, an era like the 19th Century when scientific research was uninhibited by dogmas, when electrical induction, X-Rays and electromagnetic waves were first observed and many other wonderful discoveries were made. If superluminal velocities were possible, then we might even in our time begin discovering physical effects that are unknown still because the limitations of our sensory faculties normally hide them from observation.

When the neutrino experiments at the CERN no longer showed superluminal velocities however, no signs of disappointment were recognizable anywhere in the world of science. Measurements ended as soon as they had shown a neutrino time of flight 'consistent with the speed of light'. That was the treacherous wording of CERN's press release of June 8, 2012, revealing that being right had once again been considered more important than making new discoveries. The constancy of the speed of light thus turned out to be the *Iron Curtain of Science*, strictly separating permitted research from forbidden research even in experimental physics.

The theory of relativity which had let down that Iron Curtain was the culmination point of theoretical physics in more than only one respect. It was the ultimate attempt at solving from the desktop, like God the Almighty himself by the sheer power of our brains, the great problems of physics that faced humankind in Einstein's time and still face it today. It also was the last attempt made in earnest at making the deepest desire of many theoretical physicists of that period come true, the desire for combining in physics the subjective certitude of empirical knowledge with the objective certainty of geometry, algebra and logic.

Both attempts were doomed from the outset because the goals Einstein had set himself were incompatible. Man's will can change. In fact it changes far too often. Yet with all its fickleness it still cannot have it both ways. It cannot aim at the same time at being infallible and therefore right in all circumstances, and at accepting experience as the sole and independent arbiter of science. The fact that Einstein himself and all those following him failed in their quest for a general field theory is the empirical confirmation of that conflict in the goals he had set himself.

In the context of ether theory, which is the most important topic of this *Second Part*, we must discuss different aspects of Einstein's theory of relativity. This chapter will only be about his special theory. Others will follow in *Chapter 11*.

I

Einstein first published his special theory in 1905 in his paper *Zur Elektrodynamik bewegter Körper* (On the Electrodynamics of Moving Bodies). We already saw the mathematical self-contradiction in its basic equations in the *Introduction* of this essay. I will discuss its mathematical side in more detail further down but must begin with other aspects here because they will be important for understanding the aims he was pursuing. At the very beginning of that paper he promised that

> 'introducing a "luminiferous ether" will … prove unnecessary,'

In spite of that however, he professed himself a believer in the ether hypothesis, though with limitations depriving it of all intelligible content. In 1920 he said in a speech in honour of Lorentz at Leyden:

> 'Denying the existence of ether would ultimately mean to assume that empty space is without any physical properties at all. With this view, the fundamental facts of mechanics are not consistent.
> …
> The ether of the General Theory of Relativity is a medium which is itself devoid of *all* mechanical and kinematic properties, but will

influence mechanical (and electromagnetic) events.

…

According to the General Theory of Relativity, space without ether is unthinkable'.[286] (Einstein's italics).

For trying to understand those words we must bear in mind that Lorentz was a staunch supporter of the ether hypothesis and remained so to the end of his life. Einstein, world-famous already at that time, was probably trying to be to be kind to him when he spoke those words.

One can hardly overlook though that he was speaking in enigmas even then. And besides, would an honest statement of diverging opinions not have been more ethical among scientists? My impression is that he felt more than he thought, and that he showed little sense of responsibility in framing his sentences. He did not clearly distinguish the world of physical facts from the world of human ideas. At any rate he never explained how 'a medium … devoid of *all* mechanical and kinematic properties' (the italics were his!) could nevertheless 'influence mechanical (and electromagnetic) events'. His attitude towards ether theory was ambivalent if anything, and that was symptomatic not only of him but also of the whole situation of theoretical physics in the first half of the 20th Century.

For understanding that situation we must take a closer look at ether theory. We saw it only as an attempt at solving some problems of physics so far but we have now reached a point where we can discuss it under the more general aspect of epistemology.

(*1*) It is not hard to imagine that ether theory must have seemed fishy to some scientists in the 19th Century, particularly to those cherishing an axiomatic understanding of science. It could lay claim to a certain amount of plausibility but hardly more than that. Admittedly the ether hypothesis had provided explanations for some phenomena which would otherwise

---

286 Einstein, *Äther und Relativitätstheorie* (Ether and the Theory of Relativity), Collected Papers of Albert Einstein, vol.7, p. 160..

have remained unexplained. But even the best of explanations could not obliterate the fact that it had been invented for just that. Nobody had ever been able to make ether visible. Under no conceivable circumstances could it guarantee the absolute certainty on which physical science had to build in axiomatic understanding. Ether theory relied on empirical speculation without deducing it from reliable premises. It must have been a thorn in the side of any scientist trusting in the axiomatic approach.

The fact that the ether hypothesis had been purposely designed for explaining the phenomena of visible light was apt to discredit it seriously in the eyes of an axiomatist. Inventing an entirely new substance that persistently eluded observation just for explaining some physical phenomena tasted of improvisation and kludge. It seemed 'unscientific' if not frivolous, unworthy at any rate of serious scholars. James Clerk Maxwell (1831–1879), one of the most influential physicists of his time, summarized those qualms in the words

'To fill all space with a new medium whenever any new phenomenon is to be explained is by no means philosophical, ...'

He continued that sentence, however, by saying

'... but if the study of two different branches of science has independently suggested the idea of a medium, and if the properties which must be attributed to the medium in order to account for electromagnetic phenomena are of the same kind as those which we attribute to the luminiferous medium in order to account for the phenomena of light, the evidence for the physical existence of the medium will be considerably strengthened.'

In a way Maxwell had thus staked out the boundaries which common understanding set to tolerating scientific speculation in his time. On the one hand ether theory was tainted with the stigma of being a mere *ad hoc* hypothesis not mathematically deducible from valid premises. On the other hand it nevertheless seemed just barely tolerable because it explained not only one physical phenomenon but offered a unified explanation for at least two phenomena otherwise unconnected, the phenomena of light and those of electromagnetism. It was a scientific crutch, so we will have to understand him, which science might tolerate for the time being but only as long as no alternative had been found deductively.

In the eyes of adherents of axiomatic theory the right to exist of the ether hypothesis was weak thus. It was further weakened in the years after 1887 when Michelson made his experiments for making visible the effects of ether wind. He carried them out with utmost diligence and at great expense, repeating them several times. Many scientists of the late 19th Century, especially those favouring an axiomatic understanding of science, regarded them as attempts at 'verifying' the existence of ether. One did not know then that it is logically impossible to 'verify' laws of nature that are to be valid indefinitely. Karl Popper was to show that only decades later in his *Logik der Forschung* (1934), and some theoretical physicists tend to ignore it to this day. We must bear that in mind. In Michelson's time many believed his experiments to be attempts at 'verifying' the existence of ether wind, and as such they had failed invariably. Despite greatest efforts it had never been possible to make ether visible at least indirectly by its effects. That is why some consider Michelson's experiments the biggest failure in the history of experimental physics.

Seen from the angle of modern theory of science that interpretation is untenable. Empirical theories must be open to empirical criticism. We can believe them to be true as long as no experiment refuted them, but we can never 'verify' them or 'justify' them in any clear sense of those words. We saw that in *Chapter 2, II*. The problem of ether theory is not in verifying the existence of ether but in the question of the refutability of the theory. I will discuss it in *Chapter 11*.

At this point there is no need to go into details yet because even in Einstein's time and regardless of differences of opinions on epistemology most scientists would probably have agreed that the ether hypothesis was in no way suited for serving as one of those basic axioms for which the adherents of axiomatic theory were searching. An unknown substance called 'ether' which had never been made visible directly or indirectly was in a way just the opposite of the secure foundation on which axiomatic understanding wanted to erect the edifice of science. The ether hypothesis could certainly not lay claim to being one of its fundamental axioms. On the contrary, adopting it would even have raised the question of whether such axioms could possibly exist if a completely unknown substance filled the whole universe. The ether hypothesis did not fit the axiomatic approach to science at all. Rather, it called it in question in principle.

(2) Seen before that historical background two of the famous essays published by Einstein in 1905 appear like a pincer attack on ether theory. Like Wellington and Blücher at Waterloo he turned against his adversary from different sides in separate but well coordinated attacks.

Maxwell had taken the ether hypothesis to be (just) tolerable because it offered a unified explanation for the effects of light and those of electromagnetism. Einstein's paper *Concerning an Heuristic Point of View Toward the Emission and Transformation of Light*, which we discussed in the previous chapter, dealt with the theory of light and aimed at showing how it could do without ether by assuming the existence of light quanta. And his paper *On the Electrodynamics of Moving Bodies*, containing the special theory of relativity, dealt with phenomena which Maxwell had named 'electromagnetism' and others later generalized by the name 'electrodynamics'. For those phenomena Einstein proposed a new theory not depending on the ether hypothesis, and he even promised explicitly that the introduction of 'luminiferous ether' would 'prove to be unnecessary'. In combination those two essays published within a period of only a few months aimed precisely at overthrowing the two pillars supporting Maxwell's edict of toleration, and thus at destroying the very foundation of ether theory itself.

(3) That reveals the true reason for the demise of ether theory in the first decades of the 20th Century. I am not suggesting that Einstein alone brought it down. Others before him had prepared the ground. His attack could only be successful because many distrusted ether theory in his time. It did not fall victim to the attack of any individual. Its real undoing was

the general understanding of science which had spread at the turn of the 19th Century, making demands on physical theories which *no* empirical theory could have met.

The turn of the 19th Century was the high tide of *uncritical* rationalism in the sense of Popper's theory; we must keep that in mind. His *Logik der Forschung* (1934) was a challenge to the intellectual climate of that period, and the approach he criticized there prevailed by no means in physics alone. Just over a century after Kant's *Critique of Pure Reason* (1781) the inherent limitations of rationality which he had shown there seemed hardly to bother anyone any more. Blind faith in the omnipotence of mathematics had spread widely, and logical deduction was indiscriminately identified with scientific thinking and with deductive systems. A craving for exactness appearing almost pathological from today's perspective was at large, but few seemed to realise its preconditions or to distinguish between axiomatic theories and empirical theories. In my country some believed even the legal profession could gain its results by logical deduction from legal concepts exclusively, and would become an 'exact science' by doing so[287]. That atmosphere of uncritical pseudo-rationalism also explains why Max Planck and Einstein presupposed the 'ideal' conditions of black bodies or of constant motion on straight lines in their theories. They must have thought that only ideal conditions ould reveal the mathematical principles underlying them.

After the tumultuous developments of science in the 18th and 19th Century that attitude was understandable to some degree because progress had been stupendous indeed. The whole world had been witnessing the spectacular results of physics, chemistry and medicine, watching them mostly with admiration but perhaps sometimes also with envy. Technological inventions and scientific discoveries had set off the beginning of industrialization, and their results were so stunning that many tried to emulate the natural sciences whence they originated. But not all understood the conditions to which the natural sciences owed their great success. The

---

287 This refers to Hans Kelsen's *Pure Theory of Law*. For details, see my *Recht und Rationalität*, p. 52 (footnote).

greatest misunderstanding arose in theoretical physics because the axiomatic understanding of science had spread widely there. Many were interested mainly in mathematics and were searching only for mathematical support of physical theories. They feared that physics might lose its status as an 'exact science' if it were unable to generate findings reached by pure deduction which would thus be reliably secured as they believed. The attitude was so dominant at that time that one of the rare critics of Einstein even spoke of the 'terror regime of mathematicians' in that context[288].

Ether theory could not possibly meet the expectations of exactitude raised by the axiomatic approach. The various ether hypotheses existing in the 19th Century were almost the opposite of mathematics. They originated in empirical speculation, they were highly imaginative and daring, and they were quite incalculable for the time being. All attempts at confirming empirically the existence of ether had failed, and any hope of being able one day to state the theory of ether in the precise terms of mathematics must have seemed utopian then. Not even an approach was known to describing it in mathematical equations[289]. Anyone believing earnestly in the axiomatic understanding of science could never consider ether theory to be 'scientific'.

The meal was prepared thus on which the theory of relativity could thrive and throw ether theory from the nest. Experiments might still have shown whether the ether hypothesis was good for anything but it never came to that any more after Einstein had published his thoughts[290]. There was not even a discussion on whether it still was possible to uphold the ether hypothesis, at least no audible discussion continuing until a decidable result had been reached. Instead, the whole theory was set aside unceremoniously because it did not fall in with the intellectual climate

---

288 H. Fricke, preface to *Der Fehler in Einsteins Relativitätstheorie* (1920).

289 Huygens had made an approach in his *Traité de la Lumière* (1690) but it seems to have been widely ignored.

290 Some later experiments can well be interpreted as having confirmed the existence of ether, particularly Sagnac's experiment and those on 'time dilation' made by Hafele/Keating. In *Popper versus Einstein*, I proposed other experiments for putting it to the test (Ch. 10, p. 187–195).

of the period. Einstein's attack seemed to confirm the reservations which physicists trained to the axiomatic approach had against the ether hypothesis, and Einstein promised to get on without speculation. To all those who would rather leave thinking to others his theory seemed preferable therefore. One wanted to believe him and one trusted Max Planck who agreed with him on that point. But one did not care for critical discussion.

II

Einstein's paper *On the Electrodynamics of Moving Bodies* caught attention from the outset and probably still is one of the most famous treatises in the history of science. Occasion for criticism was not lacking therefore, but the paper was widely known and seldom read. Many did not even notice that Einstein had not *deduced* in any way the constancy of the speed of light there, nor given any other reasons for believing it to be true empirically, but had merely *presupposed* it as a hypothesis. Scrutinizing the paper reveals also significant contradictions. They were inherent in Einstein's approach, they continued in the terms he used, and they culminated in his mathematical equations. Due to his elegant but somewhat sophisticated style and the high level of abstraction of his paper, and most of all due to lack of critical faculties in his contemporaries, those mistakes remained hidden to theoretical physics for a whole century.

I begin by demonstrating Einstein's mistakes only by the means of normal common sense. Discussing their mathematical and methodological aspects will follow in sections *III* and *IV.*

(*1*) The allegation that time is not 'absolute' but 'relative' is one of the key messages of special relativity. It is an implication of the so-called 'Lorentz Transformation', a set of equations used for converting the coordinates of stationary and moving systems while taking into account the principle of the constancy of the speed of light. In that context almost everybody will have heard of the implication that a traveller returning to earth after a long and fast journey through space will have remained younger than his twin brother who had stayed here. I discussed one aspect of it in *Chapter 6, IV, 3.* One of Einstein's lectures illustrates the importance he attached to that implication. He wrote there:

> 'The law of the constancy of the speed of light in empty space, as confirmed by the development of electrodynamics and optics, in conjunction with the equality of all inertial systems (special principle of relativity) demonstrated by Michelson's famous experiment, first implied that *the concept of time had to be qualified by allocating to each inertial system its proper time.*
>
> . . .
>
> According to the Special Theory of Relativity, space and time coordinates have absolute nature insofar as they are directly measurable by clocks and rigid bodies. *But they are relative in so far as they depend on the state of motion of the selected inertial frame'.* [291] (My translation and italics).

(2) The following will show that Einstein's reasoning was self-contradicting. Let a missile take off from the prime meridian at Greenwich on some random day but precisely at the moment when the sun is at the apex of its path there. By the definition of Greenwich Mean Time that would be 12.00 *h* sharp. The missile is to be equipped with propulsion systems which we control from the earth. We let it make a fast journey through space and return to Greenwich where it falls into the Thames again precisely on the prime meridian and precisely when the sun is at its apex there.

The missile will have taken exactly as long for its journey as the earth takes for one rotation on its axis. By the definition of Greenwich Mean Time that would be 24 hours. The theory of relativity claims however, that the missile carries with it its own 'proper time' which will be slowed

---

291 Einstein, *Über Relativitätstheorie, Eine Londoner Rede*, in *Mein Weltbild'*, p. 217 ff., 218 f. – The text quoted is also another example for Einstein's essentialist approach mentioned above in section *II* of the *Introduction*.

down due to the high speed of the missile. The 'proper time' of the missile at its return would therefore have to be not 12.00 h but 12.00 h minus some difference Δ (some split nanoseconds perhaps). That difference may be extremely small but there must be one because otherwise the theory of relativity would be meaningless. Yet we observe with our own eyes that the sun is exactly at his highest point again at the return of the missile, showing us that the earth has turned on its axis exactly once during the trip.

What are we to infer from this? If there really were a difference between Greenwich Mean Time and the 'proper time' of the missile as postulated by the special theory of relativity, then that difference would have to exist also at the same location. In our example it would have to exist on the prime meridian at Greenwich because the missile returned there. For us the time at Greenwich is 12.00 h but according to the theory of relativity it is to be 12.00 h minus Δ for the missile. That raises the question of what standard might apply to measuring the 'proper time' of the missile.

It cannot be the standard of Greenwich Mean Time derived from the rotation of the earth because in that case we could derive the following statements from the theory of relativity:

(i) Greenwich Mean Time at the return of the missile is 12.00 h.
(ii) Greenwich Mean Time at the return of the missile is *not* 12.00 h.

They cannot both be true because that would violate either the meaning of the word 'true' or that of the word 'not' used in statement (ii)[292]. Nor can hours, minutes and seconds serve as units for defining or measuring the 'proper time' of the missile because they too were derived from the earth's rotation. What other benchmark could apply however? *No adherent of the theory of relativity ever answered that question in more than a hundred years!* None was interested in actually measuring the time of anything. But they all believed they had understood the *essence* of 'time' and that they alone knew what time 'really is'.

---

292 I explained the definition of 'not' in *Chapter 4, II, 2.*

Some theorists of relativity are wont to refer to other definitions of the units of time in this context. Not the earth's rotation is to serve as a standard but for example the highly regular oscillations of atoms used in caesium clocks. Occasionally one even reads that such clocks are more accurate than the rotation of the earth itself.

Whatever else one may think of such proposals, they certainly do not resolve the logical problems of the theory of relativity. Using an atomic clock for reference instead of measuring time by the earth's rotation would make no difference to the logical aspects of the argument. The time units of the atomic clock would replace those defined by the earth's rotation but the conflict would remain the same. The missile would start from the prime meridian when the atomic clock resting at Greenwich shows 12.00 h sharp, and it would return from its trip when the clock again shows 12.00 h sharp. According to the special theory of relativity however, its 'proper time' would have to be 12.00 h minus Δ. Therefore the atomic clock resting at Greenwich could not provide a valid scale for measuring the so-called 'proper time' of the missile because that would lead to the same contradiction as that shown above (i, ii). It would still be necessary to find some other standard for measuring the 'proper time' of the missile.

(3) Apart from demonstrating the self-contradiction in Einstein's theory the argument also shows how little sense there is in using largely un-known physical entities such as the speed of light as benchmarks for mea-suring things better known than the benchmark itself. In the long history of physics scientists and others generally took the reverse approach, at least insofar as they were successful. They measured unknown sizes or effects by the scale of better known sizes or effects and they fared well by that method. For measuring distances they used the sizes of feet or of other parts of the human body. The well-known effects of the freezing point and the boiling point of water at sea level, or the normal temperature of the human body served as criteria for measuring temperatures. The light of a wax candle became the unit of brightness and the power of a horse became the unit of physical performance. And they measured time by the

position of the sun which meant, as Copernicus showed, that the motion of the earth relative to the sun served as a scale.

In all those cases physics started from well-known basic sizes and gradually developed from them systems of standards by introducing more reliable definitions and mathematical classifications. Even the basic parameters themselves originated in conventions. The very term 'benchmark' indicates that. The limbs of individual human beings may have different sizes but our ancestors solved the problem arising from that by agreeing on some individual foot or arm and marking its length on a bench. In every case well-known entities on which one could easily agree served as the starting points of those systems of standards.

Both Planck and Einstein however wanted to take the reverse approach. Instead of using the known for measuring the unknown they wanted to use the unknown as a benchmark for measuring the known. Planck wanted the fundamental constant h to serve as foundation of a standard system although it was not accessible to empirical observation. And Einstein wanted the speed of light, denominated V or c, to serve as the benchmark for measuring time although it had been the origin of all the problems of theoretical physics remaining unresolved to this day. Together those two scientists thus brought confusion to its peak.

III

We now come to Einstein's own presentation of his special theory and to some methodological aspects of his approaches. Once again our main topic will be the conflict between his axiomatic view of science and the heuristic approach in his thinking.

(*1*) Einstein began his paper by stating that a theory founded on the notion of absolute rest will result in asymmetries when applied to moving bodies. He based that on sound empirical arguments and then proceeded as follows:

'Examples of this sort, in combination with the unsuccessful attempts at discovering any motion of the earth relatively to the "medium of light", lead to the assumption that in mechanics as well as in electrodynamics no properties of phenomena exist which correspond to the *concept of absolute rest*, but that rather, as has been shown already for quantities of the first order, the same laws of electrodynamics and optics will be valid for all *systems of coordinates to which the equations of mechanics apply.* We will raise this conjecture (the purport of which will hereafter be called the "Principle of Relativity") to the status of a premise, and introduce also another premise, irreconcilable with the former only seemingly, namely, that *in empty space light is always propagated with a definite velocity V which is independent of the state of motion of the emitting body.* These two premises will be sufficient for attaining a simple and consistent electrodynamics of moving bodies based on Maxwell's theory for stationary bodies. The introduction of a "luminiferous ether" will prove to be superfluous inasmuch as the view here to be developed will not require an "absolutely stationary space" provided with special properties, nor assign a velocity-vector to any point of the empty space in which electromagnetic processes take place.' (My translation and italics[293]).

---

293 The most widely accepted translation of Einstein's paper probably is that by Perrett and Jeffery in *The Principle of Relativity*, published in 1923 by Methuen and Company, Ltd. of London. I have tried to be closer to Einstein's text in my translation. Readers may want to compare it with that by Perrett and Jeffery. It reads like this:

'Examples of this sort, together with the unsuccessful attempts to discover any motion of the earth relatively to the "light medium," suggest that the phenomena of electrodynamics as well as of mechanics possess no properties corresponding to the idea of absolute rest. They suggest rather that, as has already been shown to the first order of small quantities, the same laws of electrodynamics and optics will be valid for all frames of reference for which the equations of mechanics hold good. We will raise this conjecture (the purport of which will hereafter be called the "Principle of Relativity") to the status of a postulate, and also introduce another postulate, which is only apparently irreconcilable with the foœ, namely, that light is always propagated in empty space with a definite velocity *c* which is independent of the state of motion of the emitting body. These two postulates suffice for the attainment of a simple and consistent theory of the electrodynamics of moving bodies based on Maxwell's theory for stationary bodies. The introduction of a "luminiferous ether" will prove to be superfluous inasmuch as the view here to be developed will not require an "absolutely stationary space" provided with special properties, nor assign a velocity-vector to a point of the empty space in which electromagnetic processes take place.'

(2) Einstein's heuristic approach shows in the fact that he made two assumptions at the beginning of his paper and then immediately raised them to the rank of premises (or 'postulates') without attempting to justify them in any way.

The first of those premises was that 'the same laws of electrodynamics and optics will be valid for all systems of coordinates to which the equations of mechanics apply'. Transferring that to the level of empirics its purport is that the laws of nature valid here on our planet apply throughout the universe. The other premise was in his principle of the constancy of the speed of light, assuming that 'in empty space light is always propagated with a definite velocity V which is independent of the state of motion of the emitting body'.

In the context of his paper Einstein probably intended those premises to be empirical statements. We cannot be sure of that however, because he did not draw the line between empirics and mathematics as we would draw it in our days. The text last quoted shows that he used the term 'systems of coordinates' not only as an abstract of mathematics but sometimes also as standing for real space in which the laws of electrodynamics and optics are valid. The other premise 'that light is always propagated in empty space with a definite velocity V, which is independent of the state of motion of the emitting body', is empirical insofar as it contains a statement about how light will actually behave. By narrowing down that to the case of 'empty space' however, and also narrowing down to constant motion on straight lines the range of his special theory, Einstein presupposed conditions not existing anywhere in nature, at least not within reach of humankind. At that point his approach strongly resembled that of Max Planck who took the conditions of an 'ideal' black body as the starting

---

I am not altogether happy with that translation because using the word 'postulate' for empirical statements, such as statements about the speed of light, is misleading in my opinion. The German term used by Einstein was 'Voraussetzung'. In the context of his paper I would translate that by 'precondition' or by 'premise'. In fact, I think the only really clear terms would be 'hypothesis', or 'conjecture' but we will see later that using them in the translation would be clearer than Einstein's thinking had been. Perrett and Jeffery used 'conjecture' in the context of the principle of relativity but not in that of the constancy of the speed of light.

point of his calculations. It therefore remains questionable whether Einstein's premises really deserve the epithet 'empirical'.

Einstein neither tried to deduce those premises from others statements nor to justify them in any other way. They were to convince all by themselves through the results following from them. If they had been empirical in the sense of Popper's theory, they might have been called 'conjectures' or 'hypotheses' that would have to be probed by making targeted experiments.

However, the sentences just quoted also indicate that Einstein did not intend to sacrifice his axiomatic creed on the altar of his heuristic approach. On the contrary, the fact that he placed those statements about the 'concept of absolute rest' and about 'systems of coordinates' at the beginning of his paper already foreshadows that he was seeking the solution to his problems in formulae and figures rather than in physical phenomena. It also indicates that he did not distinguish clearly the physical *explanation* of such phenomena from their *description* in terms of mathematics. Since that confusion is quite common still in our days, it is likely that Einstein did not realise that the words 'explanation' and 'description' usually have different meanigs, in common language as well as in the language of science[294].

(3) Einstein's partiality for mathematical solutions shows most of all in what he did *not* discuss in his papers *On the Electrodynamics of Moving Bodies* and *Concerning an Heuristic Point of View Toward the Emission and Transformation of Light*. In both papers his omissions deserve at least as much attention as his statements.

The issues covered by those papers concern among other things the theory of light. We saw in *Chapter 7* that it had relied on the ether hypothesis since the times of Huygens and Newton. Einstein's aim was directly opposed to that. He wanted to show that 'the introduction of a "luminiferous ether" will prove to be superfluous'. From a conscientious

---

294 For my definition of those terms for the purpose of this essay, see section *II, 1* of the *Introduction*, above.

scientist one should therefore have expected that he would devote particular attention to the phenomena of light. After all, Rœmer's discovery that the speed of light is finite had been the decisive event convincing so many scientists of past centuries that light must have a carrier and that the ether hypothesis must be a true description of reality. And the discoveries of polarization and interference had been why it had gained even more power of persuasion in the 19th Century.

Nothing of that kind happened however. Einstein had not mentioned those physical phenomena in his paper *Concerning an Heuristic Point of View Toward the Emission and Transformation of Light*[295], and he not even hinted at them in his paper *On the Electrodynamics of Moving Bodies*. He did not discuss how we might explain the propagation of light without a carrier. He neither mentioned Young's double slit experiment which had led to the discovery of light interference, nor the phenomenon of polarization which also suppors the wave theory of light. Huygens' and Faraday's explanations of the phenomena of light in analogy to sound got no mention in either paper. All the important discoveries of the 19th Century relating to light and radiation remained undiscussed. Anyone inclined to disbelieve that should read those papers again under this aspect.

Suppressing the most important counter-arguments to a theory proposed in a scientific presentation would normally be considered unreliable and even disreputable. Whosoever wants to protect Einstein from that accusation can only refer to his unconditional belief in the axiomatic method. I at least can think of no other excuse for an account which withheld *all* the most important counter-arguments known at the time. But this one reveals the limits that were set him.

---

295 In that paper (American Journal of Physics, v. 33, n. 5, May 1965) the only reference to such phenomena was at the end of the second paragraph where Einstein wrote 'In spite of the complete experimental confirmation of the theory as applied to diffraction, reflection, refraction, dispersion, etc., it is still conceivable that the theory of light which operates with continuous spatial functions may lead to contradictions with experience when it is applied to the phenomena of emission and transformation of light.' - He did not mention interference and polarization.

At the time of publishing those papers in 1905 Einstein was a young scientist, aged only 26 years. One of his teachers at Zurich may have put him on the wrong track. And reading Maxwell, his idol, may have added to his confusion as we will see further down in this chapter[296]. At any rate he must have lost himself in the clutter of mathematical theorems and physical theories and phenomena, and his contact with the empirical side of physics must have broken down almost completely. That sometimes happens to young academics. The surprising aspect of those developments is that contemporary theoretical physicists did not notice his blunders and instead of criticizing him styled him a genius.

My own interpretation is different nevertheless. I think anyone looking for *explanations* in Einstein's papers and expecting from him critical discussions of physical phenomena has misunderstood him. He did not even try to explain the constancy of the speed of light which he presupposed. Nor did he try to explain the effects of interference and polarization of light. In fact, I think he did not try to explain anything, at least not in those papers. Least of all did he want to make new empirical discoveries. For in that case he would have had to explain the known by the unknown as we saw[297]. He would thereby have contradicted diametrically his own views of physics as an 'exact science'.

I think all Einstein wanted to achieve in those papers was setting up mathematical approaches for *describing* the phenomena of light. Following Maxwell's example his aim was to express in precise mathematical formulae what he belieed to be the certain knowledge owned by physical science, and to confine his endeavours strictly to that. Having realised that Maxwell's equations lead to 'asymmetries' when applied to more than one body moving relative to another within one system of coordinates[298], he

---

296 I will discuss Maxwell's understanding of electricity in the context of Einstein's formula $e = mc^2$ in section *VI, 1*.

297 See Popper, *Realism and the Aim of Science* (1983), p. 132. I quoted his text in section *II, 1* of the *Introduction*.

298 He mentioned that in the first sentence of his paper *On the Electrodynamics of Moving Bodies*. I presume he was referring by that to Newton's *Hypothesis I*, quoted in *Chapter 7* at the end of *sec. I*.

thought the solution could be found by turning the constant motion of time into a variable and the variable motion of light into a constant. He therefore sought to set up the mathematical equations needed for calculating the motion of physical bodies in space under those assumptions. Since he could not establish by deduction the initial operands on which his equations had to rely, in particular that of the constant speed of light, he raised them to the rank of premises. They were the axioms from which he started. Like Max Planck he believed them to be true because calculation would have been impossible without them[299]. He did not notice the conflict between a method gaining its basic statements by deduction from axioms that had been precisely defined and the method of trial and error which gets along without deducing its hypotheses because they will have to be probed afterwards by the empirical standard of experience. At any rate it did not seem to bother him.

That is the only explanation I can offer for the fact that in his papers *Concerning an Heuristic Point of View Toward the Emission and Transformation of Light* and *On the Electrodynamics of Moving Bodies,* Einstein left undiscussed *all* the counter-arguments known at the time. The fact itself is undeniable.

IV

The problems of Einstein's special theory arose from his assumptions about the speed of light. For observing the rules of logic he should either have given up the assumption that the speed of light is constant or that it is independent of the motion of its source. Otherwise he would inevitably end up in formulae containing also mathematical contradictions[300]. I con-

---

299 For my interpretation of Planck's view see *Chapter 8, II, 2.*

300 The implication is important because it assigns to mathematics its proper place in the theory of knowledge.
Its necessity follows from the fact that mathematics itself is a language, only expressed in shorter symbols. We can re-translate all expressions of mathematics into normal language

fess however that I realised this implication only after publishing *Popper versus Einstein* (1998). In the light of later developments that warrants a short digression.

In *Popper versus Einstein* I had had been content with demonstrating in common language the logical mistakes and conceptual muddles in Einstein's theories of relativity. My strongest logical argument against them was the one explained in *Section II, 2* above. Their mathematical side seemed unimportant to me then because I knew that a theory based on contradicting premises could never yield valid inferences. That is a simple truism implied in the definition of 'truth'. If the premises on which we base an inference contradict one another, and if we have accurately determined their meaning in the respective context, then not all of them can be true. Therefore the rules of inference cannot guarantee the truth of conclusions drawn from them. Instead, we can infer from them with seeming logical stringency any conclusion whatsoever. All results inferred from contradicting premises are *random*. That is universally recognized in mathematics as well as in logic. It can even be *proved* in the strictest sense of the word[301]. Theories based on contradicting premises are useless in science because the results inferred from them must always be arbitrary.

That was the state of my knowledge at the time of writing *Popper versus Einstein*. Surprisingly however, I had to realise that those few theoretical physicists with whom I could speak or correspond in person would be quite unimpressed by arguments at the level of so-called 'formal' logic even when discussing a theory as formal as Einstein's special theory claims to be. They either were unacquainted with the truism that contradictions in the premises of a theory must *necessarily* lead to results that are arbitrary or were unwilling to accept its implications with respect to the special theory. My logical arguments were set aside and I was repeatedly told

---

again, but the result would be long and complicated sentences more difficult to understand than mathematical formulae.

301   Popper, among others, demonstrated that proof in *What is Dialectic?* (*Conjectures and Refutations* p. 317–319).

that my knowledge of mathematics was insufficient for understanding the special theory.

During that process I began to realise that the logical mistakes of the special theory of relativity which I had criticized already in *Popper versus Einstein* must necessarily reappear also in Einstein's mathematical formulae. That caused me to search for mathematical contradictions specifically in his paper, admittedly with some apprehensions due to my limited knowledge of mathematics, but they turned out to be unfounded. Finding the mathematical self-contradiction shown in the *Introduction* of this essay was not difficult because I knew what I was looking for. Nevertheless it still seems almost incredible to me that a mathematical contradiction as obvious as the one I have shown could have remained undiscovered in a whole century. Yet the fact that no textbook on special relativity mentions it seems to confirm that.

Having found Einstein's mathematical mistake did not improve my situation however. Previously I had been told that my knowledge of mathematics was insufficient for understanding his special theory. From then on I was accused of being unable to understand his texts. And after I had published the mathematical proof of Einstein's self-contradiction in 2012 there never came any reaction to it from the side of theoretical physics[302]. That confirmed me in my view. If anybody could say something against that proof, then surely some theoretical physicist would have done so by now.

Having explained that history of my endeavours I will now discuss in more detail the mathematical approaches of Einstein's special theory of relativity.

(*1*) Einstein claimed that his theory rested only on the two premises mentioned above: the premise that

> 'the same laws of electrodynamics and optics will be valid for all
> systems of coordinates to which the equations of mechanics apply',

---

302 C.v.Mettenheim, *The Oscillation Project with Emulsion-Tracking Apparatus (OPERA) experiment: An argument for superluminal velocities?* Physics Essays, vol. 25 (2012), p. 397–403.

and the premise that

'light is always propagated in empty space with a definite velocity V which is independent of the state of motion of the *emitting*
body'. (My italics).

The first premise already indicates his confusion of mathematics and physics. The laws of electrodynamics and optics are of laws of nature which we
believe to be true. They cannot be 'valid' in systems of coordinates. We
can only try to use systems of coordinates for describing them.

We can leave it at that however because the second premise is more
interesting. It is usually called the 'principle of the constancy of the speed
of light' but it has two aspects which we must discuss separately. One of
them is that the vacuum speed of light is to be always the same, and the
other is that it is to be independent of the state of motion of its source. The
latter aspect is the origin of all the difficulties which non-physicists still
experience with relativity to this day. We must examine it closely because
there is some truth in it. Yet it also is the point where the shift of meaning
in Einstein's reasoning came in, a shift of the same type as that in my
Thatcher example given above, only more subtle.

(*a*) The assumption that the speed of light is independent of the state of
motion of its source expresses for light only what we already know about
sound from the Doppler principle discussed in *Chapter 6, II, 2*. After sound
separated from its source, not the motion of that source but only the physical properties of the transporting medium, usually air, will from then on determine its speed of propagation. That was what Einstein wanted to assume
also for light, though without referring to any medium transporting it. My
italics in the text last quoted show that. Up to that point his premise is well
in line with physical experience concerning sound.

Unfortunately Einstein did not stop there. He proceeded by transforming his premise into mathematical equations, and unnoticed by him
those equations presupposed something different from what he had stated
in his premise. From there the drama unfolded.

(*b*) Einstein was trying to set up equations for calculating distance and speed of physical bodies while taking into account that the light signals used for measuring would not be instantaneous but would travel at the finite speed of light. How could we measure the simultaneousness of events under those conditions? That was the question from which he started. For setting up mathematical equations he needed precisely defined operands. In § 1 of his paper he therefore began by defining the simultaneousness of two events occurring at the space coordinates *A* and *B* of a system of coordinates with clocks at *A* and at *B*. To that effect he wrote:

'In order to render our presentation more precise and to distinguish this system of co-ordinates verbally from others, which will be introduced hereafter, we call it "system at rest".'[303]

His ensuing definition of simultaneousness thus referred explicitly to the space coordinates of a *system at rest*. That will be important for further discussion. In the same section he wrote but a little further down:

'However, it is not possible without further stipulations to compare, in respect of time, an event at *A* with an event at *B*. We have so far defined only an "*A* time" and a "*B* time." We have not defined a common "time" for *A* and *B*, for the latter cannot be defined at all unless we establish *by definition* that the "time" required by light to travel from *A* to *B* equals the "time" it requires to travel from *B* to *A*. Let a ray of light start at the "*A* time" t_A from *A* towards *B*, let it at the "*B* time" t_B be reflected at *B* in the direction of *A*, and arrive again at *A* at the "*A* time" t'_A. According to our definition the two clocks synchronize if

$$t_B - t_A = t'_A - t_B \qquad [304] \qquad (1)$$

303 Albert Einstein, *Zur Elektrodynamik bewegter Körper* (On the Electrodynamics of Moving Bodies), Annalen der Physik vol. 17 [1905], p. 891 ff. The translation is from '*The Principle of Relativity* published by Methuen & Co. (1923).

304 In equation (1), t_A stands for 'time at *A*', t_B for 'time at *B*', t'_A stands for for 'time at *A* after reflection of light in *B*'.

We assume that this definition of synchronism is free from contra-dictions, and possible for any number of points; ...

In agreement with experience, we further assume the quantity

$$\frac{2\,\overline{AB}}{t'_A - t_B} = V \qquad (2)$$

to be a universal constant (the speed of light in empty space).'
(Einstein's italics).

This leaves no doubt that Einstein's definitions of simultaneousness in equation (1) and of the speed of light in equation (2) *explicitly referred to the system at rest*. Further down in that paper however, he transferred them to the moving system unchanged.

(2) Before coming to that we must look at equation (2) more closely be-cause that is where Einstein's mistake came in.

Equation (2) is a definition of V standing for the speed of light. In the numerator of the left side $(2\,\overline{AB})$ Einstein doubled the distance from A to B, and in the denominator he used the expression $t'_A - t_A$ as standing for the time difference between the emission of the light signal in A and its return to A after reflection in B.

(a) It would have been simpler to use only AB in the numerator and only the time taken by light on the single track from A to B in the denominator. The equation would then read

$$\frac{\overline{AB}}{t'_A - t_B} = V \qquad (2a)$$

As a definition of the speed of light in a system *at rest* equation (2a) would have been algebraically as correct as equation (2). However, equation (2a) would have been a definition of the speed of light on the single track from A to B whereas Einstein's equation (2), using double track in the numerator and dividing it by the time taken on both tracks in the denominator, in fact defines V by the *average* time taken by light on two tracks.

In a system at rest that makes no difference. Even in a moving system we could neglect the proper speed of the system because that in one direction will compensate that in the other. *But that only holds for speed, not for distance.*

(*b*) For seeing that, we must now look at some physical aspects of the situation Einstein had in mind. He may have believed the difference between defining the speed of light on one track or on two tracks to be irrelevant but in that case he would have been mistaken. The formula $t_B - t_A = t'_A - t_B$ (1) is correct only if light takes the same time on its way from A to B and on its way back from B to A. That is self-evident in a system at rest but in a moving system it is not true.

Another look at the transport of sound will show that. Let an ambulance A drive at high speed, using its siren while approaching the observer B standing at the roadside. The time taken by the sound signal for transmission of from A to B is independent of the speed of the vehicle. Once sound impulses have left the siren and hit the air and caused it to vibrate, the vehicle will no longer influence the transmission of sound. Only the air transporting sound will do so. Even if the vehicle were to stop, or to turn off, or to disintegrate into its component parts, that would influence the propagation of sound no more than destroying the runway of an airport after take-off of an aeroplane will disturb its flight. Sound is on its way, the air is transporting it and it no longer has anything to do with the source from which it came.

The same does not hold for a sound signal sent from B to A however. If the observer B, standing at the roadside, sends a sound signal to the approaching ambulance A, then the time of that transmission may be independent of the speed of sound but it is not independent of the speed of the ambulance, because in that case the ambulance at A is not the source of the signal but its receiver. While the signal is travelling the vehicle is approaching and thereby shortening its path from B to A. Assuming the velocity of the signal to be the same on both ways, as Einstein did, the motion of the vehicle will shorten *also the transfer time* of the signal. Einstein's formula (1) $t_B - t_A = t'_A - t_B$ is true for sound only if it is travelling in a system at rest but certainly not for sound travelling in a moving system. Then why should it be true for light?

The following diagram shows that the lengths of the ways of light are also different.

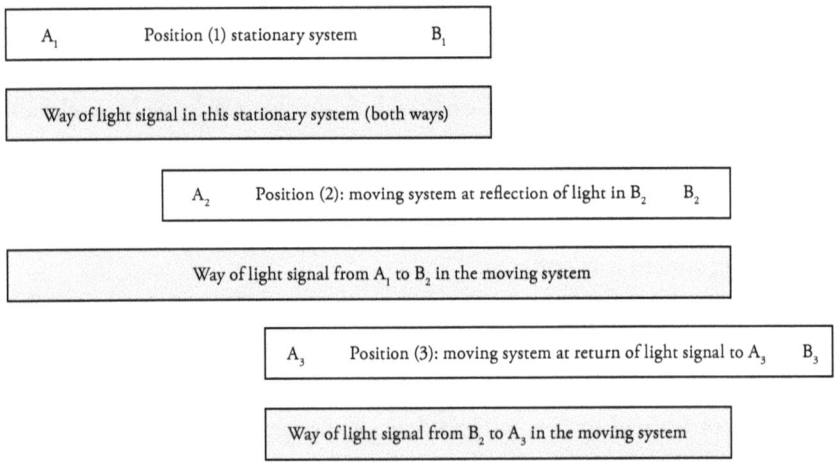

A₁   Position (1) stationary system   B₁

Way of light signal in this stationary system (both ways)

A₂   Position (2): moving system at reflection of light in B₂   B₂

Way of light signal from A₁ to B₂ in the moving system

A₃   Position (3): moving system at return of light signal to A₃   B₃

Way of light signal from B₂ to A₃ in the moving system

Diagram: Light signal starting in A is reflected in B while the system AB moves from position (1) to position (3).

The way of the light signal from A₁ to B₂ obviously is longer than that on its return from B₂ to A₃. Even if the speed of light in a vacuum really were constant, independent of the motion of its source, Einstein's equation (1) would still not apply to a moving system.

(c) It seems that Einstein never saw that; at any rate he did not discuss it in his paper. In the next section, § 2 headed 'On the Relativity of Lengths and Times', he transferred his equations (1) and (2) unchanged to a moving system AB at the ends of which there were again to be clocks. He wrote there:

'We imagine further that with each clock there is a moving observer, and that these observers apply to both clocks *the criterion
established in § 1 for the synchronization of two clocks*. Let a ray of
light depart from A at the time t_A, let it be reflected at B at the time
t_B, and reach A again at the time t'_A. *Taking into consideration the
principle of the constancy of the velocity of light* we find that

$$t_B - t_A = \frac{r_B}{V - v} \tag{3}$$

and [305]

$$t'_A - t_B = \frac{r_B}{V + v} \tag{4}$$

where r_{AB} denotes the length of the moving rod measured in the
stationary system. Observers moving with the moving rod would
thus find that the two clocks were not synchronous, while observers in the stationary system would declare the clocks to be
synchronous.

So we see that we cannot attach any "absolute" signification
to the concept of simultaneousness, but that two events which,
viewed from a system of co-ordinates, are simultaneous, can no
longer be looked upon as simultaneous events when envisaged
from a system which is in motion relatively to that system.' (My
italics).

This text shows that Einstein did not act unintentionally when he trans-
fered his equations (1) and (2) to the moving system AB. On the contrary,
he even gave specific reasons for that transfer. By referring to 'the criterion established in § 1 for the synchronization of two clocks' he *explicitly*
pointed to equation (1). His 'taking into consideration the principle of the

---

305   In equations (2) and (3), Y_{AB} stands for 'length of the moving system', V for 'velocity
of light' (explained in Einstein's text to equation [1a] quoted above), and v for 'speed of
the moving system'.

constancy of the velocity of light' could only mean that this principle was to apply also to the moving system. Otherwise that statement would have been meaningless.

Any doubts remaining after that will be removed by a footnote which Einstein made on page 896 of his paper for explaining equations (3) and (4). It reads like this (with my italics for text put in quotation marks by Einstein):

> 'Zeit bedeutet hier *Zeit des ruhenden Systems* und zugleich *Zeiger-stellung der bewegten Uhr, welche sich an dem Orte, von welchem die Rede ist, befindet*'.

> In English: '*Time here* stands for *time of the system at rest* and also for the *position of the hands of the moving clock which is at the place under discussion*'.

He could have expressed himself more clearly of course[306]. Nevertheless the words 'und zugleich' ('and also') leave no doubt that the meaning of t and t' in his formulae (3) and (4) was to be the same as that in the system at rest which is that of formula (1). He also said explicitly that he was applying this identical concept to the system at rest *and* to the moving system. Otherwise his equations (3) and (4) would have been undefined.

(*3*) The self-contradiction in Einstein's reasoning becomes visible if we compare his equation (1) with equations (3) and (4). We already saw that in the *Introduction* of this essay but I must repeat it here for clarity's sake.

The expressions on the left sides of (3) and (4) are the same as those on both sides of (1). By the rules of algebra we may therefore insert the right sides of (3) and (4) in (1)which gets us

---

306 Giving the same meaning to the terms 'time' and the 'hands of a clock' confuses time with an instrument for measuring it, as we saw in the discussion of time dilation in *Chapter 6, III*. Einstein's paper may have been the origin of that mistake which modern textbooks still repeat.

$$\frac{r_A}{V-v} = \frac{r_B}{V+v} \tag{5}$$

Eliminating the identical numerators on both sides reduces that to

$$V - V = v + v \qquad \text{or} \qquad +v = -v \tag{6a, b}$$

which is equivalent to

$$v \neq v \qquad \text{or} \qquad + = - \text{ (plus equals minus)} . \tag{6c, d}$$

That *proves* the self-contradiction as well as the shift of meaning located in v. The only case in which equation (5) could be valid would be $v = 0$ which is the case of the system at rest. If applying equations (3) and (4) were limited to that case, then they would no longer be *functional* equations describing the relationship between distance and time of moving systems at *varying* speeds. They would apply only to static systems. That in turn would raise the question of what point there might be in having a theory about the relativity of time applying only to motionless systems in which time is no relevant parameter.

(4) Einstein's mistake consisted in shifting the meaning of his terms by transferring to the moving system the definition of simultaneousness given in equation (1). That shift happened in equation (2). His premise had been

> 'that light is always propagated in empty space with a *definite* velocity V which is independent of the state of motion of the *emitting* body'. (My italics).

When defining V through the speed of light on two tracks in equation (1a) however, he tacitly shifted the meaning of that premise to saying

> 'that light is always propagated in empty space with a definite ve-

locity V which is independent of the state of motion of the emitting *or the receiving* body'.

Einstein must have overlooked that shift of meaning in equation (2) because otherwise he could never have reached his contradicting equations (3) and (4). That has nothing to do with his premise of the constancy of the speed of light as yet. It merely results from the fact that due to the proper speed of the moving system, the *distance* from A_1 to B_2 is different from that from B_2 to A_3. The diagram above shows that. Einstein overlooked the fact that the reflection of light in B implies a change of roles because from then on B is the source of light and A is the receiver.

(5) In spite of all those facts one physicist argued that in formula (1) Einstein had defined simultaneousness only for stationary clocks, but did not *want* that definition to apply also to formulae (3) and (4) which he intended only for moving systems. That was the only counter-argument I ever read[307].

The argument is incompatible with Einstein's main text and with his footnote quoted above. It would also leave the speed of light in the moving system undefined because in equation (2) Einstein had defined it only for the system at rest. Failing other indications I cannot believe that argument to be a true statement of Einstein's intentions at the time of writing his paper.

The argument is also invalid. Even if Einstein had really 'wanted' formula (1) to apply only to stationary systems and formulae (3) and (4) to apply only to moving systems, that would neither remove the contradiction in his equations nor the invalidity of his deductions. That is why the

---

307 The physicist putting forward that argument was Markus Pössel, member of Albert Einstein Institute at Potsdam and co-responsible for the 7th ed. of Max Born, *Die Relativitätstheorie Einsteins* (2003). In a letter to the German Federal Ministry of Science written in 2005 he conceded that Einstein's equations (1), (3) and (4) are contradicting, but he contended that Einstein had been *aware* of that contradiction. He did not say how he could have come to know such particulars of Einstein's actual knowledge in 1905. That is why I think he only had difficulties in admitting his own error.

argument is also unfair to Einstein's memory. It cannot rescue his special theory of relativity, yet accuses him of dishonesty by contending that he had been aware of the problem but had decided to leave it unmentioned.

(6) The upshot of all this is that Einstein had based his Special Theory of Relativity on a shift of content as shown in my Thatcher example[308]. That shift resulted in a mathematical mistake in his equations. His derivation of the Lorentz Transformation rests on the self-contradiction shown in equations (5, 6a,b,c,d). It is mathematically invalid[309].

There are strong empirical arguments for believing the speed of light to be independent of the motion of its source, the analogy to sound probably being the strongest. But there are no arguments for believing it to be constant. Assuming the speed of light to be constant relative to the emitting body *and* relative to the receiving body would imply the self-contradiction shown above. Assuming it to be constant *only* relative to the emitting body would remove that self-contradiction but it would disagree with observations, most of all with those of the redshift[310]. Given the present state of our empirical knowledge, the only alternative is assuming the speed of light to be independent of the motion of its source but not constant. Q. E. D.

V

Some readers may be discontent at this point because after all that logic and mathematics, and though having been told that time is not relative

---

308 See *Chapter 7, III,* above.

309 As I explained at the beginning of section *II, 3,* the name 'Lorentz Transformation' in the theory of relativity stands for a set of equations used for converting the coordinates of stationary and moving systems assuming the speed of light to be constant.

310 If the speed of light were constant relative to every emitting source throughout the universe, then there would be very high differences of velocities in the light reaching us from other celestial bodies. As we saw in *Chapter 7, II,2,* that would clash with the observation that the redshift in the light coming from distant stars is comparatively small.

but absolute, they still do not know what time 'really is'. They may think that something as important as time must exist in reality, and may want to learn more about it. Even accepting all that has been said, they may still want to know whether time goes only in one direction or can run forward and backward. I am not sure that I can satisfy their needs but I think I can say where they go wrong.

My suggestion is to take the nominalist approach and begin by clearly saying what we intend to talk about. If language is manmade and if we ourselves give their meaning to the words we use[311], then we ourselves must decide what we want some word to designate, and must even explain that if necessary. We need not always do so explicitly of course. In the case of the word 'time' explaining its meaning will normally be unnecessary because the context in which we use it will tell others whether we are speaking of the time of the day, or of the time of the year, or of Greenwich Mean Time, or of the time for doing something, or whatsoever else. In any case however, we must *first* know what we are talking about and *then* give it a name for making a long story short.

All those believing in the reality of time however, and trying to find out what it 'really is', insist on doing it the other way round. They begin by uttering the word 'time' and then wonder what it might mean. That is the essentialist approach of Aristotelianism criticized so strongly by Popper in his *Open Society*[312]. He showed there that there never is any point in asking 'what is'-questions or in inquiring into the 'essence', or the 'nature', or the 'true meaning' of some concept, at least not in science, because that approach almost inevitably leads into vicious circles. That even applies to questions which pupils put to their teachers. Phrasing them in 'what is'-form may be quite common in daily life. But pupils asking questions like 'what is redshift?' will usually want to learn something quite different. They will not expect deepest insights into the 'nature' or the 'essence' of redshift. Far more likely is that they either want to find out what the term is to designate in some specific context, and will therefore be content with

---

311 See *Chapter 4, II*, above.
312 Popper, *The Open Society and its Enemies* (1945), vol. 2 ch. 11, II (p. 9–21).

getting a short *description* of the phenomenon. Or they may want to get an *explanation* of the phenomenon which would then have to consist in giving them information on its causes. Situation and context of a question will usually tell which of the two is required. The more important some word is in our language, the more different meanings it is likely to have. That makes out its importance but it also means that we may have to clarify in the respective context in which sense we are using it.

Experience shows however that those accustomed to taking the essentialist approach will feel left alone rather if given no more than the clever advice to do things the other way round. It is difficult to get out of that because the word 'time' can have so many different meanings. For those not content with defining it by the earth's position relative to the sun or by some other physical process, but preferring to search for a 'deeper' meaning and believing that to be an important philosophical issue, I have a simple suggestion for taking the nominalist approach. In that specific context they could use the word 'time', or expressions such as 'time is passing' or the like, as designating that our world does not stand still because even if seemingly nothing is changing in one place, there will still be physical processes at almost every other place of the universe. The word 'time' would then stand for the fact that physical processes are everywhere, that matter is a physical process, that we human beings are physical processes as long as we live and our remains even beyond that for a while, and that the entire universe is no static state but a dynamic process. It would express that even if things seem to be standing still in one place, processes are nevertheless going on everywhere else and can be compared with one another. All that, I suggest, can be expressed by the simple word 'time' if we use it in that 'deepest' sense. But we must be aware that there is no way of measuring time in that 'deepest' sense because for doing so we would have to compare it with *all* those other processes going on in the universe which is obviously impossible. And if there is no way of measuring time in that 'deepest' sense, then there can also be no point in using it as an operand in physical equations.

Whosoever is unwilling to accept my suggestion and still insists on discussing the 'true meaning' or the 'nature' of time should now make a counter-proposal. For I believe, as Karl Popper did, that 'to utter a word and mean nothing by it is unworthy of a philosopher'[313].

Those claiming to be following my suggestion however, but contending nevertheless that 'time can run backwards' somehow, would first have to explain what the word 'time' is to stand for in that specific context. They particularly would have to clarify whether they are contending that all physical processes are running backwards in time or just one individual process in relation to other processes. If they want to say that only the Earth's rotation changes direction I would have no theoretical problems but the more doubts about the empirical truth of their allegation. If they claim, however, that only some individual particle or body is 'moving backwards in time', then I would argue that they are contradicting themselves by going back on their own words. The meaning of the concept 'time' proposed above, on which we fictitiously agreed, did not say anything about forward or backward movements. It only said that that our world does not stand still because at any point of the universe there will be changes even if seemingly nothing is changing at some specific place. In our fictitious agreement the word 'time' thus stands only for the fact that physical processes are everywhere, that is to say for the situation that matter is a physical process, that we human beings are physical processes and that the entire universe is not a static state but a dynamic process. Assuming *individual* processes to be running forward or backward in time is incompatible with that very general understanding of the concept 'time' because we would then first have to agree on what we mean by such a direction.

I would therefore request my fictitious interlocutors to tell me once more what meaning they want the word 'time' to designate when they use it in that 'deepest' sense. Since no one ever explained that in literature my personal opinion is that those deeming possible that individual particles could move *backwards in time* know not even what they are talking about.

---

313 Popper, *Conjectures and Refutations*, p. 166.

The chapter could have ended here if Einstein had not presented his famous formula $e = mc^2$ as an afterthought to his special theory of relativity[314]. He published it also in 1905 and it is reputed to have established the *principle of the equivalence of mass and energy* in physics[315]. In *Chapter 6* we saw that adherents of relativity later interpreted it as having explained or even predicted nuclear fission more than three decades before Hahn and Strassmann actually discovered it. Popular representations of the theory of relativity hardly ever omit it. So we must discuss also $e = mc^2$.

(*1*) Einstein's paper *Ist die Trägheit eines Körpers von seinem Energieinhalt abhängig?* (hereinafter: '$e = mc^2$-paper') holds only three pages. At its end he reached the following conclusion:

> 'The mass of a body is a measure of its energy-content; if the energy changes by L, the mass changes in the same sense by $L/9 x 10^{20}$, the energy being measured in *ergs* and the mass in *grammes*.'

Literally taken that not only means that two kilogrammes of coal contain twice as much energy as one kilogramme of coal but also that one kilogramme of water contains as much energy as one kilogramme of coal. To my mind that is an empirical refutation of $e = mc^2$. At the very least it shows that as long as we cannot burn water the formula is inept for solving any of the practical problems of energy facing mankind. Besides, Einstein claimed to have inferred it by strict deduction from the special theory of relativity. The self-contradiction underlying that theory therefore is also at

---

314 Albert Einstein, *Ist die Trägheit eines Körpers von seinem Energieinhalt abhängig?* (Does the Inertia of a Body Depend on its Energy Content?) Annalen der Physik Bd. 18 (1905), S. 639.) Annalen der Physik vol. 18 (1905), p. 639. - In the German original of that paper Einstein used a different notation.

315 A.P. French, *Special Relativity*, (ed. 1997). p. 16–18.

the bottom of $e = mc^2$. All inferences drawn from it are random for that reason alone[316].

We might leave it at that if the validity of $e = mc^2$ alone were at stake, but his paper is interesting for other reasons. It shows that his difficulties arose from his concept of 'energy', and points to their origin in Maxwell's theory. My own impression is that at the time of writing the $e = mc^2$-paper, Einstein must still have been completely entangled in the self-contradictions of special relativity and having greatest difficulties with understanding energy at all. He must also have been firmly convinced that the laws of conservation were given by nature itself.

(a) The most obvious mistake in Einstein's $e = mc^2$-paper consists in a *petitio principii*. He was presupposing in his premises what he was trying to prove in the conclusion. Showing that is fairly simple.

Logical and mathematical operations are tautological in the sense that a conclusion drawn by correct inference cannot contain information that was not already given in one of its premises. We saw that in *Chapter 2, III*. In the $e = mc^2$-paper the word 'mass' does not appear anywhere in the premises of special relativity from which Einstein started, but only in the conclusions he drew from them. That alone already indicates that he cannot have reached those conclusions by correct inference. He must have shifted the meaning of his terms in the process of deduction. Scrutinizing his paper confirms that surmise.

Einstein introduced no less than four different symbols for designating 'energy' in that paper (L, H, E and K), each of them in two varieties standing for energy in the moving system and for that at rest. In his denomination K stands for 'kinetic energy', and he defined H and E like this:

> 'H and E are energy values *of the same body* referred to two systems of co-ordinates which are in motion relatively to each other, the body being at rest in one of the two systems.' (My italics). `

---

316 For details, see *Chapter 8, IV.*

By those words Einstein was presupposing that the energy of a physical body would vary, depending on the system of coordinates from which it is being considered. The self-contradiction underlying his special theory therefore was also underlying his deduction of $e = mc^2$, but that is not all. In those definitions Einstein was also presupposing that there is such a thing as an *inherent* energy of physical bodies which is neither identical with ponderable mass nor with calories. Otherwise there would have been no need for assigning two different energy values to one body irrespective of its specific components, particularly not if that body is 'at rest in one of the two systems'.

The assumption that energy is inherent in physical bodies shows also in Einstein's first introduction of E_0, which he did in the following words:

> 'Let there be a stationary body in the system (x, y, z), and let its energy referred to the system (x, y, z) be E_0.'

A stationary body can have no kinetic energy. If that statement is to make any sense therefore, that is to say if the body is to be stationary in the system (x, y, z), yet have the energy E_0 in that same system, then E_0 must be standing for something else than kinetic energy, and there is no other possibility than assuming it to stand for inherent energy. Einstein therefore *presupposed* that there is such a thing as an inherent energy of a body even when at rest, and he did not attribute that energy to any specific components of that body. His equations were to apply to coal and water alike.

Several other instances confirm this interpretation. In his paper *On the Electrodynamics of Moving Bodies* Einstein repeatedly spoke of 'electric mass' ('elektrische Masse')[317]. His translators even had difficulties in finding adequate English terms for that expression, which is understandable since electricity is unknown to have any mass still in our days[318]. At the

---

317 Annalen der Physik vol. 17 [1905], p. 891 ff., § 6 (3rd sentence on p. 907 of the German original).

318 In the context of physics, the German word 'Masse' can only stand for 'mass'. The term 'elektrische Masse' must therefore be translated by 'electric mass', whatever that may mean. The translation by Perrett and Jeffery in '*The Principle of Relativity* (Methuen &

beginning of his paper *Concerning an Heuristic Point of View Toward the Emission and Transformation of Light*, published earlier that year, he wrote:

> 'The energy of a ponderable body cannot be subdivided into arbitrarily many or arbitrarily small parts, *while the energy* of a beam of light from a point source (according to the Maxwellian theory of light or, more generally, according to any wave theory) *is continuously spread* (over) an ever increasing *volume*.' (My italics).

Further down in that paper he repeatedly used the term 'volume' for designating *quantities* of energy or radiation[319]. When introducing K as standing for 'kinetic energy' in the $e = mc^2$-paper he did not explain what other kinds of energy the other symbols were to designate, in particular E_0. All this shows that he was treating electricity like a physical body, somewhat in the way of bulk cargo. If electricity were a physical body, then that body would of course have to consist of energy. His final inference that

'the mass of a body is a measure of its energy content',

thus was presupposed already in his first premise assuming bodies to have inherent energy even when at rest and irrespective of their specific components, be they coal or water. His deduction of that inference was circular.

(*b*) The more interesting question is how Einstein came to believe that there is such a thing as inherent energy in a body at rest and what he meant by it.

Co., 1923) mostly does so. In section II, § 6, however, in the section beginning with the words 'As to the interpretation of these equations . . .' ('Zur Interpretation dieser Gleichungen . . .') they translate 'elektrische Masse' by 'electric charge'. Retranslated into German that would be 'elektrische Ladung'.

319   He wrote, for instance: 'Suppose we have *radiation occupying a volume* v' (p. 6); or: 'The following equation applies when the temperature of a unit *volume of blackbody radiation* increases by dT' (p.7; my italics).

The answer to that is to be found in Maxwell's theory of electrodynamics to which Einstein explicitly referred in both papers. For understanding the situation we must bear in mind that the theory of electricity still was in its infancy when Maxwell wrote his *Treatise on Electricity and Magnetism* published in 1873, and that it had raised entirely new problems. In our time, when electricity has become a permanent companion in almost every minute of our lives, it hardly is conceivable anymore how strange and miraculous its effects must have seemed to those having spent the greater part of their lives without it. That must have been Maxwell's situation too. With static electricity he got on fairly well in the first volume of his book although he defined it as a mathematical *quantity* even there, and tried to describe it in terms of 'density'[320]. But in the second volume he had greatest difficulties when discussing electric current. The only way he could find for visualising it was by comparing it to a *fluid*. But his problem was that this view did not agree altogether with other phenomena, such as static electricity, or batteries, or the accumulators then recently invented. The best way of showing his difficulties is by quoting him verbatim. He wrote there[321]:

> '*The energy of an electric current is either of that form which consists in the actual motion of matter, or of that which consists in the capacity for being set in motion.*
>
> The first kind of energy, that of motion, is called Kinetic energy, and when once understood it appears so fundamental a fact of nature that we can hardly conceive the possibility of resolving it into anything else. The second kind of energy, that depending on position, is called Potential energy, and is due to the action of what we call forces, that is to say, tendencies towards change of relative position.
>
> ...
>
> *The electric current cannot be conceived except as a kinetic phenomenon.*
>
> ...

---

320   James Clerk Maxwell, *A Treatise on Electricity and Magnetism*, (1873), Oxford-edition (1998), vol. 1, p. 71, 72.

321   Maxwell, *A Treatise on Electricity and Magnetism*, vol. 2, p. 211, 213.

We therefore know enough of electric currents to recognise in a system of material conductors carrying currents, a dynamical system which is the seat of energy, *part of which may be kinetic and part potential.*' (My italics).

Those sentences show that with electric current Maxwell was all at sea[322]. He saw but two ways of explaining it. It had to be either *kinetic* energy 'which consists in the actual motion of matter' (!) or *potential* energy. And since neither seemed to fit altogether, his conclusion was that both must be true though both only partly. And he never explained how that could be.

Maxwell based his further inferences on those assumptions. For explaining the propagation of energy through electric current he relied on kinetic energy which he visualized as resembling the transportation of a liquid. The fact that this view could not explain all aspects of electric current did not seem to worry him.

(*c*) Apparently Maxwell did not know that Huygens, living two centuries earlier (1629–1695), had already shown that nature has a way of transporting energy without transporting matter itself. I already mentioned his theory in *Chapter 7, I,* but we need more details at this point.

---

322 Maxwell was not alone with that in his time. Even the *names* for designating the new parameters of electric tension, current and resistance had to be created first. Ohm seems to have been the first to use the word 'tension' for designating potential electric energy (Georg Ohm, *Versuch einer Theorie der durch galvanische Kräfte hervorgebrachten elektroskopischen Erscheinungen*, Annalen der Physik und Chemie, vol. 82 [1826], p. 459–469). Maxwell however, still used the same word in a quite different meaning in his *Treatise on Electricity and Magnetism* (Oxford-edition [1998], sections 48 and 241). The next problem consisted in finding reliable definitions for those parameters, and in measuring them without disturbing them. For solving that, Wilhelm Weber and my ancestor Rudolf Kohlrausch invented an instrument for measuring electric current not through electrolysis, which was the only way known then, but through its magnetic effect (Weber/Kohlrausch, *Über die Elektrizitätsmenge, welche bei galvanischen Strömen durch den Querschnitt der Kette fließt*, Ann.d.Phys. vol.XCIX [1856], p.10–25).

In his *Traité de la Lumière* (1690)[323] Huygens had assumed ether to consist of smallest elastic particles standing in immediate contact with one another, and had thus explained the phenomena of light by direct influence of ether particles on adjacent ether particles. Due to their high elasticity and immediate contact light would propagate in spherical waves analogous to the propagation of sound. He demonstrated the possibility of that by the example of a number of balls lying in a straight line and in direct contact with one another. An impulse received by the first ball will pass on to the last ball and only this ball will then be propagated whereas the balls between the first and the last will remain almost stationary. Thus his theory did not explain the high speed of light by any particle *travelling* as fast as light but only by the *elasticity* of ether particles[324]. The high velocity with which they react to an impulse coming from adjacent particles by contracting and re-expanding would explain the high velocity of light.

Huygens' notion that transporting energy need not imply transporting matter itself revealed a general principle more important even than his explanation of the phenomena of light. Newtonian mechanistic doctrine can describe energy only as manifesting itself in the *motion* of material bodies or particles, and both Maxwell and Einstein became victims of that limitation. In Huygens' theory however each individual ether particle will keep its place and only change its shape for a short moment by elastically responding to an impulse received. Yet the elastic properties of the particles would carry the energy impulses of sound or of light over great distances, and not the *resistance* of the transporting medium but only its *elasticity* would determine the speed of propagation. Energy is no transport of anything material in Huygens' theory but a state of excitation of the transporting medium, leading thus to a fundamentally different view of related phenomena.

Huygens' theory is true in the sense that it describes a way of transporting energy that actually exists in nature. There is no need to revert

---

323 Huygens wrote his *Traité de la Lumière* in 1678 and read it to the British Royal Society in that year, but published it only in 1690.

324 Huygens, *Traité de la Lumière*, p. 14–18.

to ether theory or to the theory of light for being convinced of that. The well-known device shown in *Figure 4* below is an excellent demonstration of the underlying principle. It is usually known as 'Newton's Cradle' but I will call it 'Huygens' Cradle' from now on because that name is more appropriate. Anyone playing at billiards will also know that a billiard ball can pass on to the adjacent ball the energy of an impulse received without noticeably changing its own place.

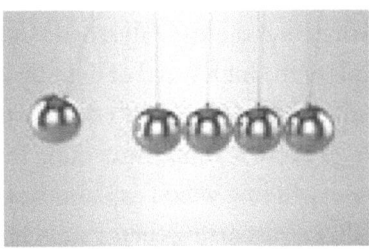

Figure 4: The ball on the left will hit the second ball, and only the last ball on the right will bounce off.

Huygens showed all that in detail, and the great tsunami of December 26, 2004 was a sad demonstration of the principle he had revealed[325]. The energy of the undersea earthquake in Indonesia reached the coast of South Africa within only a few hours, causing serious damage even there. According to newspaper reports it propagated at a speed of about 1000 km/h. No one will believe even a single water molecule to have travelled so fast through the Indian Ocean. The only possible explanation is that not water but pure energy alone had travelled that distance at such speed. Neither Maxwell nor Einstein however considered that possibility in their time although Huygens had explained it very clearly.

For physical theory there is an obvious lesson to learn from that. If nature actually uses the method of transporting energy without transporting matter itself when the medium is as crude as water, then no reason is con-

---

325 Huygens, *Traité de la Lumière*, p.14.

ceivable why we should assume the propagation of other kinds of energy, such as electricity or light, to consist in *transporting* something material.

(2) It is hard to believe that Maxwell could have overlooked Huygens' theory but that seems to be how it was. For trying to understand that we must go back even to the times of Huygens and Newton.

In 1671 Newton had published his famous *New Theory About Light and Colors*, showing that the white light of the sun is composed of rays of different colours[326]. In 1673 there was a short correspondence between him and Huygens on that paper. In 1675 Newton sent another paper to the Royal Society, this time bearing the title *Hypothesis explaining the properties of light*. That title leaves no doubt that he was speculating in this case, and that he knew he was. His normal attitude was different. In his scientific publications he *described* the physical phenomena he was investigating and confined himself strictly to that. In his paper of 1675 however, he departed from that principle and by using the word 'Hypothesis' made very clear that he was doing so. He tried to *explain* the phenomena of light there through the hypothesis of its consisting of particles traversing ether.

That paper was published only in 1757 however, when both Huygens and Newton were long dead[327]. Huygens therefore knew Newton's *New Theory About Light and Colors* (1671) when he wrote his own *Traité de la Lumière* in 1678 and read it to the *British Royal Society* of which he was a fellow[328]. But he seems not to have known Newton's later *Hypothesis explaining the properties of light* because it had not yet been published. Newton was a fellow of the *Royal Society* since 1672 and its president from 1703 to his death in 1726. It is hardly conceivable therefore that Huyges' theory of the propagation of energy could have escaped his notice. But in his *Prinicipia*, published in 1687, he returned to his usual attitude of not discussing mere hypotheses. That must have been why he never men-

---

326 Isaac Newton, *A New Theory About Light and Colors*, Philosophical Transactions 6 (1671), p. 3075ff, 1078.

327 Thomas Birch, *The History of the Royal Society*, vol. 3 (1757), pp. 247–305.

328 In 1673 there was a short correspondence on that paper between Huygens and Newton. Huygens mentioned that in the paper published in 1690.

tioned Huygens' theory in his *Principia*, nor in any other publication that I know of[329]. He discussed energy but the only kinds he mentioned were kinetic energy and ponderable energy.

The result of that situation seems to have been that Maxwell, in his boundless admiration of Newton, believed Newton's *Principia* to contain all that could be said about light. And Einstein blindly followed Maxwell on that path. Both were more interested in mathematical descriptions than in physical explanations, and they never wondered about how light propagated in reality.

Some of this is guesswork of course, but it is the only explanation I can offer for the fact that Huygens' explanation of the transportation of energy remained undiscussed by Newton, Maxwell and Einstein.

(3) As a result of all that, Einstein was left with the problem of having to describe the propagation of energy as the motion of particles and therefore in categories of *kinetic* energy. His method for solving that problem in the $e = mc^2$-paper was the same as that employed in his light quantum hypothesis. He shifted the meaning of his terms by simply equating mass and energy as we saw above.

Other than Max Planck however, Einstein was not describing something he already knew, but was trying to solve a real problem of physics. Mass and energy are in fact interchangeable to some extent. Photosynthesis will show us every summer that the energy of the sun can turn into mass in the shape of trees, and our fireplace will show us every winter that the wood of trees can turn into energy again. Assuming some law of nature to be governing that process is reasonable therefore.

Einstein however was not content with that, and again it was his belief in the axiomatic approach that led him astray. Instead of searching for a specific law of nature solving only that specific problem, he presupposed that there must be 'general elementary laws from which, by pure deduc-

---

329 In a letter to Boyle, dated Feb. 28, 1678, Newton discussed a variety of the ether hypothesis coming quite close to that proposed by Huygens. Cajori quotes from that letter in *Sir Isaac Newton's Mathematical Principles of Natural Philosophy and his System of the World*, vol. II, pp. 675.

tion, the worldview can be gained'. Pure deduction however presupposed fixed operands, and fixed operands were possible only if the total of energy in the universe always remained the same. The interchangeability of mass and energy then probably led him to the conclusion that there must be not only a law of conservation of energy but also a corresponding law of conservation of mass, so that the total of energy and mass in the universe always remained the same. Firm belief in the laws of conservation was no result of the axiomatic understanding of science therefore. It was its most essential *precondition*. One believed their validity to be self-evident. Not even Karl Popper called that into question[330], and I admit that only working on this book made me see things differently.

I no longer consider the laws of conservation as even *claiming* to empirical truth now. That would be sophomoric because we have no reason to believe we know anything about the total of mass and energy in the universe, neither that it is constant nor that it varies. There neither is nor ever will be a way of putting that to the test empirically. Nevertheless *assuming* mass and energy to be constant seems to be the only way of making some phenomena calculable for us. And making them calculable is the only way of finding out more about them by inventing theories and putting their more remote consequences to the test in experiments.

In my view the laws of conservation are *instruments of method* therefore. They are empirical statements because they refer to physical reality. But they could never satisfy Popper's criterion of falsifiability because there is no possible way of knowing whether or not the total of mass and energy in the universe remains the same. The only way out is to consider the laws of conservation as instruments for getting as near to truth as we can. If we do so however, then we must bear in mind that the basic assumptions from which we start are at best only approximations to truth.

An analogous principle might apply also to the constancy of the speed of light. At the end of section *IV* of this chapter we saw that it does *not* remain the same. Yet we need mathematical operands for applying the

---

330 Popper, *Quantum Theory and the Schism in Physics*, p. 92, 189–192; *The Open Universe*, p. 171.

rules of geometry and mathematics to physical theories about the phenomena of light. There seems to be no other way of designing experiments for testing their more remote consequences. But it would be a serious fallacy to consider those operands exempt from discussion or even to be true empirically just because we need them. If we question them however, then we must remember that the standard for gauging them is not the standard of truth but that of expediency.

However all that may be, one of the mistaken assumptions in Einstein's $e = mc^2$-paper was that *all* kinds of matter can turn into energy. As long as we measure energy in *erg* or *J*, that is in terms of its power of achieving physical performance here on earth, not even the discovery of radioactivity and of nuclear fission can convince me that one kilogramme of water contains as much energy as one kilogramme of coal.

CHAPTER 10:

WHAT ABOUT THE ATOMS? ON THE USE OF MODEL THEORIES

,Se non è vero, è molto ben trovato.'

GIORDANO BRUNO

Returning to ether theory inevitable but difficult in present state of theoretical physics; requires reconsidering from scratch. I. Planetary model of the atom never justified deductively; several competing theories but no empirical evidence; planetary model of the atom the most plausible; II. Parallelism to planetary system need not be given up but can even be made stronger by returning to ether theory; development of atomic theory from Rutherford to Bohr; quantum mechanics based on pure speculation attempting to connect quantum theory with periodic system; untenable after breakdown of quantum theory; no contribution to empirical discoveries. III. Planetary model of the atom must be given up: discovery of too many particles changed the problem situation; some discoveries by experiment, some by inference; planetary model defeated its own ends; historical reasons of atomism: problems of change and of simplicity; stages of development; setback by discovery of elements and electrical charge; planetary model of the atom would need explanation itself; situation comparable to end of Ptolemaic theory. IV. Planetary model of the atom feigns knowledge where none exists; sophomoric belief in being near to limits of nature; ultimate limits of nature may be much further removed.

From the point of view of the progress of physical theory Einstein's attack on the ether hypothesis was a failure. He had wanted to show that 'the introduction of a "luminiferous ether" will prove to be superfluous'. Instead, his light quantum hypothesis, his special theory of relativity and his formula e = mc² purporting the equivalence of energy and matter, all proved to be mistaken. Falling back on ether theory will be inevitable if theoretical physics is ever to make progress again.

For seeing things that way it is essential to accept logical arguments if they are valid. Even in science that attitude is not self-evident. It presupposes a personal decision of the scientist, recognizing that he himself is responsible for what he takes to be the truth. He must answer for his own diligence in accepting theories or designing them himself, for his readiness to admit empirical evidence when testing them, and even for his willingness to recognize and observe the rules of logic. Logic is not an expression of truth but it can be an expression of veracity. In daily life other standards often apply. Scientists too may shirk responsibility; some may even consider personal interests more important than sound arguments.

Whatever may have been the reasons, in practice at any rate Einstein's attack on the ether hypothesis was highly successful. Despite the logical mistakes he made, it contributed decisively to discrediting ether theory so thoroughly that for more than a century hardly anyone took it seriously as a physical theory any more. At present it can barely assume more than the role of a historical reminiscence. Any theoretical physicist venturing in our time to campaign for an ether hypothesis, however it be phrased, would probably jeopardize his own career. Structures are so encrusted by now that only outsiders like myself can take that risk.

Even for outsiders however the question arises whether there still is any hope that the situation could mend. The odds are against that because inertia is a powerful force, not only physically but also in the institutions of science. Much would be gained already if theoretical physicists would at least discuss the relevance of logic, and how anyone could claim to be searching for empirical truth without observing its rules. In such a discussion there would at least be a chance of some scientist mustering the courage needed for independent thinking. But even that is faced with almost insurmountable difficulties at present.

Most of those difficulties are rooted in the axiomatic understanding of science. It dominates theoretical physics still in our days, and its psychological effects are more damaging even than the logical mistakes underlying it. Many of those trusting in it have developed a tendency to shirk open-minded discussion. In their scientific education they were taught to believe that science must convey unassailable truths and that a scientist must be able to recognize them. Being unable to live up to that standard in practice, they have developed a way of twisting the meaning of their concepts until they ostensively fit the facts, and some believe that to be the true method of science. Niels Bohr actually argued that contradictions are not objectionable in science but could be overcome by introducing the method of *dialectics* and by interpreting contradictions of physical theories as a 'dualism' of physical phenomena. Heisenberg agreed with him, and both were applauded for their deep insights[331]. That may have been where theoretical physics touched bottom but Planck's introducing 'probability considerations' for reaching the results desired was not much better[332]. Physicists trained to that tradition will tend to maintain at least an outward appearance of incontestability even if inwardly they feel by no means sure at all. They will shirk discussions when they do not feel up to them, and break them off when arguments fail them. That happened to me more than once.

The actual problem of theoretical physics is even greater however. If there is ever to be any progress again in the theory of light and matter, then falling back on ether theory will be inevitable because only that will permit making discoveries explaining the known by the unknown[333]. But resuming ether theory requires not only rethinking some individual physical notions. It means taking an entirely new approach to the problems of light and matter, an approach that will have to clash with almost every major theory developed in the 20th Century. Physical science must review from scratch all the assumptions which today's theoretical physicists uncritically presuppose, and must call them into question without exception. I can only demonstrate that by one example here, that of the *planetary model of the atom.*

---

331 Bohr explained his view in *Atomic Theory and Description of Nature*, vol. I; also in his papers *Can Quantum-Mechanical Description of Physical Reality be Considered Complete?*, Phys. Rev. 48 (1935), pp. 696 ff.; and *On the Notions of Causality and Complementarity*, Dialectica 1948, pp. 312 ff. I discussed it in *Popper versus Einstein*, p. 161–166. – For Heisenberg's view, see his *Physics and Philosophy* (1958), Chapters II, III.

332 See Planck's text quoted in *Chapter 8, II, 1.*

333 This refers to the text quoted above in the *Introduction* (sec. *II,1*) from Karl Popper, *Realism and the Aim of Science* (1983), p. 132.

I

The planetary model of the atom seems to be recognized universally in our days. It has long become part of the worldview not only of physicists but also of the public. Contemporary theoretical physics links it with quantum mechanics and with the so-called Standard Model of Particle Physics. Whosoever dares to take on a theory of that consequence can hardly look for applause. At least in that respect Max Planck was right when he resignedly observed that

> 'a new scientific truth does not usually enforce itself by its opponents being convinced and declaring themselves enlightened, but rather by the fact that opponents gradually die out and the growing generation will be familiar with truth from the outset.'[334]
> (My translation).

Truth may not be recognizable as clearly as Planck presupposed by those words but the best hope of finding it lies in open-minded and critical discussion. In our days few seem to realise that the planetary model of the atom is but a hypothesis. It shows a *possible* way of visualizing the inner structure of matter. Before believing it to be describing empirical truth we must scrutinize it thoroughly[335]. Having already placed myself between all stools by my criticism of quantum theory and the theory of relativity, I may as well take that next step too. I think giving up the planetary model of the atom will be inevitable if theoretical physics is ever to make progress again.

(*1*) Rutherford's theory of the planetary model of the atom was a gratifying exception among the major physical theories developed in the 20th

---

334 Max Planck, *Vorträge und Erinnerungen* (1933), p. 13.

335 Heisenberg seems to have been one of the few seeing that in his time. At least in his paper *Über quantentheoretische Kinematik und Mechanik* (Mathematische Annalen, vol. 95 [1926], p. 683f., 684) he explicitly used the term 'Quantenhypothese' (quantum hypothesis).

Century. It was an imaginative *invention* and there never was any pretence of its being based on deduction, let alone on foundations believed to be unassailable. Even among theoretical physicists believing in axiomatic systems there simply was no question of all that. Instead, speculation was permitted to run wild from the outset. That is why the theory of the planetary model of the atom justly deserves having dominated more than a century of physical science. Even now the only objection against it from the point of view of scientific method is that it explains nothing beyond itself.

The permissive attitude of theoretical physics towards the planetary model of the atom is partly explicable by its exceptional charm, arising from the fact that it proposed to fall back on analogous models for explaining macrophysical and microphysical phenomena. The order ruling our solar system, as discovered by Copernicus and confirmed by Newton, was in principle to be found again also at the sub-atomic level. The enticing simplicity of that parallelism endowed the theory with a vividness endearing it to everyone, and gave countless inspirations to science.

Another reason why the theory of the planetary model of the atom could gain success so easily was in the fact that nothing was known of the structure of atoms anyway. Several competing hypotheses had been proposed in 19th Century. One of them assumed atoms to be like balls, and electrons to be like springs fixed to their surface. Another imagined electrons to be distributed randomly inside the atom like the raisins in a plum pudding. But none of them could claim preference over others by reason of arguments that would have been decidable at the time.

The theory of the planetary model thus thrust into a theoretical vacuum as it were[336]. On the one hand a theory was urgently needed because empirical evidence, most important Mendelejew's discovery of the periodicity of atomic weights (1869), seemed to indicate that atoms really existed. On the other hand no one had yet conceived an adequate hypothesis for visualizing them convincingly. Any theory filling that gap with

---

336 A similar phenomenon can be observed with respect to Big Bang theory in our days. There is no indication for its being true, and no new evidence has been found in many decades. Nevertheless, it gradually became to be considered self-evident by many scientists.

even moderately plausible conjectures was most welcome for that reason alone. And besides, nobody would have known anyway how to probe even indirectly by experiment a theory about the inner structure of the atom. That still is impossible in our days, even with the so much improved experimental means we have at our disposal. At least I myself never heard of any physical experiment carried out with the intention of putting to the test the hypothesis that an atom really consists of a nucleus orbited by electrons. Even the best of electron microscopes would not have sufficient resolving power for making that structure visible. Still less could the electrons, assumed to be so much smaller than the atom itself, be observed with microscopes using electrons for observing them.

At the beginning of the 20th Century the situation was even more desperate. There seemed to be no other way simply than by relying on the theoretical approach of the theory appearing most plausible, and that happened to be the theory of the planetary model of the atom. The danger that someone would disprove it was negligible at that time. That was how the breakthrough of the theory came about. The question now is whether that will still do after the revival of the ether hypothesis.

II

Scrutiny will reveal an indissoluble conflict between the ether hypothesis and the theory of the planetary model of the atom, and that means that only one of them could be true.

When investigating that issue we must keep two aspects in mind. One of them is that there never was any empirical confirmation for assuming the atom really to consist of a nucleus orbited by electrons. Giving up the planetary model need not get us in conflict with any empirical evidence already known. Few seem to remember that in our time but Heisenberg still saw it[337].

---

337 Heisenberg, *Über quantentheoretische Kinematik und Mechanik*, Mathematische Annalen vol. 95 (1926), p. 684, where he, correctly, spoke of the ‚quantum hypothesis‘

The other aspect is that giving up the planetary model need not imply giving up also the idea of a parallelism in the principles of order in the universe and those on the subatomic level. On the contrary, that parallelism will become even far stronger and simpler by adopting a new variety of the ether hypothesis.

(1) For realizing the deficiencies of the planetary model of the atom we must look at the sequence of events from a different angle.

In 1911 Ernest Rutherford (1871–1937) had proposed a model of the atom with electrons orbiting the nucleus in the manner of planets[338]. In 1913 Niels Bohr postulated that the torsional impulse of those orbits must be calculable as an integer multiple of Planck's constant h divided by 2π[339]. That was the origin of quantum mechanics.

Bohr however had neither based his postulate on empirical knowledge nor on mathematical deduction. It was pure speculation which he must have believed permissible in this case because it contributed so significantly to the overall harmony of the theoretical system. Heisenberg saw that clearly when he wrote[340]:

> "It was also clear to me very quickly that all of Bohr's results about the periodic system had been gained simply by intuition. Bohr had not calculated at all. He had not even tried to address those dreadful mathematical problems, but simply to interpret, vividly and conceptually correct, experience as it existed; and we know in hindsight that his theory of the periodic system was right in all material respects." (My translation).

---

and assumed the basic 'postulates' of quantum theory to be 'analogous' ('sinngemäße') descriptions of reality.

338 Rutherford, *The Scattering of* ☒ *and* ☒ *Particles by Matter and the structure of the Atom*, Philos. Mag. 21 (1911), p. 669–688. Nagaoka Hantarō had proposed the principle of the model in 1904, but Rutherford's experiments lead to a fundamental change of the proportions of the nucleus and the electrons.

339 Bohr, *On the Constitution of Atoms and Molecules*, Philos. Mag. 26 (1913) Part. I, pp. 1–25, 22–24.

340 Heisenberg, *Erinnerungen an die Entwicklung der Atomphysik in den letzten 50 Jahren* (1968), in *Deutsche und jüdische Physik* (1992), p. 187, 190.

Bohr's interpretation of quantum theory came closer to Planck's own interpretation than Einstein's had been. I mentioned that in *Chapter 8, IV.* At least in the context of quantum mechanics, Bohr no longer endowed the quantum with physical properties of its own as Einstein had done, but returned to Planck's notion that h must be an abstract mathematical operand. In his theory an atom will emit or receive energy by its electrons jumping from inner to outer orbits or *vice versa*, and he assumed Planck's constant h to be relevant for the distance of that jump. By that surmise Bohr tried to establish a link between quantum theory, the planetary model of the atom and the periodicity of atomic weights[341].

(2) Bohr's version of theory of the planetary model of the atom is untenable in any case. That has nothing to do with falling back on the ether hypothesis as yet. It follows from the fact that quantum theory itself was based on the logical mistakes shown in *Chapter 8*. Planck's quantum of action $h = 6.548 \times 10^{-27} erg \times sec$[342] is no fundamental constant of nature. Even its indivisibility never has been or will be established, and the notion of an instantaneous quantum jump consists in a self-contradiction as we saw there in *Section IV.* No reason is conceivable why Bohr's theory should nevertheless be relevant for the orbits of electrons or of other particles assumed to be existing in physical reality.

There is no reason to mourn the demise of Bohr's theory either, even if the whole of quantum mechanics goes down with it. Quantum mechanics relied on h being a universal constant. If that is not true empirically, then quantum mechanics too simply feigned a knowledge not existing. Having an operand like h may be helpful for making calculations. But that alone will not carry the inference that nature must consist of constants and that h is one of them.

History confirms that view. Bohr's postulate that the torsional mo-

---

341 Bohr, *On the Constitution of Atoms and Molecules*, Philos. Mag. 26 (1913) Part II, pp. 476–502.
342 Planck, *Vorlesungen über die Theorie der Wärmestrahlung*, p. 162; due to a change of the standards of measurement iti is now said to be $h = 6,626 \times 10^{-34}$ *Js*.

mentum of the orbits of electrons must be calculated as an integer multiple of Planck's constant h divided by 2π never contributed to any empirical discoveries of science. It never served for *explaining* anything but only for *describing* things wrongly believed to be known already[343]. Integrating Planck's mathematical quantity h in the atomic model may have gratified intellectually some scientists believing in it, but it never yielded any information that might have interested anyone else. It did not even permit reliable mathematical approaches because multiplying or dividing the torsional impulses of the orbits of electrons by 2π would have presupposed knowing at least the diameters of those orbits, and whether they were circles or ellipses. Some may believe that Bohr had established a connection between quantum theory and the periodic system. In fact however, physical theory was adapted *afterwards* in each case to findings made previously in other fields and by other means. That applies also to the new elements discovered in the 20th Century. Discovering them was not owed to Bohr's new version of the planetary model of the atom but to the regularity of periodic repetitions which Mendelejew and Meyer had discovered in 1869 and had set down in their tables of the elements. All Bohr's theory achieved was interpreting as being in harmony with the planetary model what others had previously discovered. Abandoning that theory will not hinder the progress of science therefore, but only remove a mental obstacle standing in its path.

III

Returning to ether theory will require more than only giving up quantum mechanics however. The planetary model of the atom itself already carries in it the germ of destruction. We must even give up the notion that an atom consists of a nucleus and electrons orbiting it. That will be necessary

---

343 For the distinction between *descriptions* and *explanations*, see section *II, 1* of the *Introduction,* above.

because in spite of its vividness the planetary model of the atom never contributed in any way to the progress of science. Even the problem it once was to solve has changed long since.

When Rutherford stated his theory in 1911 the world of physics looked different from the way in which theoretical physicists see it today. The atom was believed to consist only of the positively charged nucleus and one or more electrons carrying negative charges. Even the neutron was still unknown. It was 'discovered' only in 1921 after Rutherford himself had given the decisive hint. The nucleus itself was still believed to be indivisible then. Nuclear fission was only discovered in 1938. Rutherford had based his theory on the notion that the composition of matter was very simple and that its basic components were known. There was an indivisible atomic nucleus and there were electrons orbiting it, that was all.

The situation we face today is fundamentally different. The number of components of which the standard model theory assumes the atom to be composed increased continuously since Rutherford's time, and that development was unleashed by discoveries made by experimental physics. We cannot disregard that when weighing the pros and cons of ether theory.

The cloud chamber invented by Charles Wilson in 1912 made visible traces of water vapour condensation. He interpreted them as showing the paths of particles carrying electrical charges, and that started a never-ending series of discoveries of more and more new particles which all had to be housed somewhere within the atomic nucleus or its shell. Every newly discovered type of radiation was interpreted as having been caused by new particles, and therefore required including them in the model of the atom or its nucleus.

Mathematical deduction also contributed to that development. The 'discovery' of the neutron entailed those of the positron and the neutrino; and muons, pions, kaons and many more components followed them. In the theory of the nucleus there was no ending of that development after Otto Hahn had discovered nuclear fission (1938). At present physicists hardly speak of individual particles any more, but of large families of baryons and mesons consisting in turn of quarks or leptons or whatever else they may be called, to say nothing of the different spin states some of those particles are supposed to assume.

There would be no point in going into further details here. The goal of discovering ultimate, uniform and simple components of matter is far more remote at present than it seemed to be in Rutherford's time. Every single one of those 'discoveries' of new particles must correctly be considered as *refuting* the planetary model of the atom, at least in its original understanding. Only lack of imagination can explain the persistence of the theory in our days. Irrespective of returning to ether theory one thing is certain even now. By the development just shown theoretical physics not only left the path of classical atomism. It even lost sight of the goal atomism had been pursuing. The planetary model of the atom defeated its own ends.

(*1*) Classical atomism had probably started from the problem of change, arising from the ancient philosophical question of how a physical body can change and yet remain the same.

Heraclitus' river is the most famous example, but the fact that human beings or animals will age and yet retain their personality is probably the one we understand best. The notion of physical bodies being composed of smaller particles that change their positions without the body itself losing its identity seemed to open an approach to solving that problem but it soon led to the next one, this time consisting in an antinomy. Going by all one knew, every physical body was divisible. But it seemed inconceivable that this process of division could continue *ad infinitum* without the components of that body ceasing to be physical bodies themselves. That approach seemed to end in a contradiction in terms.

The philosophers of antiquity pondered deeply over this, and they thought that surely there must be an end somewhere to dividing physical bodies, and a point be reached at which no further division is possible. The imaginary last and indivisible component of matter which would then exist was named άτομοσ, the indivisible. The atoms of current theory thus actually bear their names unjustly. Not only are they far from being indivisible; they have also long forgone any claim to being the ultimate components of matter.

(2) Classical atomism did not originate from the problem of change alone however. It was also the expression of a fundamental principle of methodology which contemporary theoretical physicsists seem to have given up.

From antiquity to the present day searching for ultimate indivisible elements of matter had always been borne by the desire to reduce the phenomena of physics to as few as possible immutable units. Its goal was *simplicity* and it did not stand alone in aiming at that. Searching for immutable and irreducible units of nature and explaining physical phenomena by making them as few as possible has ever been a basic need of science. Einstein explicitly endorsed it when he searched for the ,irreducible basic elements' of nature and believed 'the supreme goal of all theory' to consist in making them 'as simple and as few as possible'[344]. A modified version of that same approach was even at the bottom of Max Planck's idea of the 'quantum of action'. Only simple solutions which our minds can grasp will satisfy our need for explanations.

Atomism was an attempt at explaining matter by such simple principles. Empedocles had known four elements, water, earth, fire and air. Aristotle added to them ether as a fifth, and that was where matters rested for a long time. Descartes even reduced the number of elements to only three as we saw in *Chapter 7, I*. The discoveries of different chemical elements made in the 17th and 18th Centuries thus were serious setbacks for atomic theory. Instead of diminishing the ultimate components of matter they considerably increased their number at first. In the further course however, they led to the discoveries of Dalton's law (1801) and of the periodic table of the elements (1869) explaining the differences between the elements not by exterior properties but by their atomic weight. The outlines of a new and simpler principle seemed to be getting in sight again.

That principle however became questionable even before it had been fully investigated. It clashed with Faraday's discovery of electrolysis (1834) which showed the need of distinguishing between atoms and their electrical charge. Even atoms of the same element could have different prop-

---

344 Einstein, *On the Methodology of Theoretical Physics*. I quoted the text in *Chapter 1, III*, above.

erties. That in turn entailed the invention of the theory of the electron by Hendrik Antoon Lorentz (1895). By distinguishing between the atomic nucleus and the electrons as carriers of electric charge he tried to reconcile atomic theory with the phenomena of electricity. But the theory of the electron was an invention nevertheless.

That simple atomic theory, admitting only of the nucleus and the electrons orbiting it, was the state of affairs which Rutherford tried to fortify by his theory of the planetary model of the atom. Niels Bohr followed him, hoping probably that assuming quantum jumps between the orbits of electrons would permit somehow to fit into that system also the various kinds of invisible rays discovered in the 19th Century.

(3) Going by the present state of science those approaches failed not for epistemological reasons alone. The large number of different particles discovered or believed to have been discovered in the 20th Century even cast a doubt on the goal itself. If all those particles did exist, and if they really did turn out to be indivisible, then current theory would not satisfy our need for explanation. Instead it would be in need of being explained itself.

There are too many particles; that is the *crux* of particle physics in its present state. Their number recently even increased again through the discovery of the Higgs boson, purported to be so important for completing the standard model of particle physics although it still leaves gravity unexplained. Seen from the angle of our need for explanations, particles have long assumed the role which the elements or the atoms had played in older theories. In fact, they are more complicated than elements or atoms had ever been. If there really were as many different particles as current theory assumes, and if science really were to advance human knowledge somehow, then we would now have to investigate the material of those particles and the causes of their different characteristics. We would have to search for a new theory at a deeper level.

The planetary model of the atom failed not in detail but in principle, even if no theoretical physicist seems to have realised that as yet. What sense could there be, after all, in having an atomic model that cnnot be tested experimentally if it can be sustained only at the cost of assuming the existence of a large variety of different particles with different properties and in different states, all of them in need of explanation themselves? If that model is so complex therefore that no one can understand it as a whole. According to present theory the atomic nucleus alone is far more complicated than Rutherford had believed the already too complicated atomic shell to be. Besides, the planetary model of the atom has not been yielding connections to reality for a long time any more. It cannot explain even gravity, the most obvious among the forces of nature. In fact, it does not explain any physical phenomena at all but only embodies the reminder of a cherished idea.

Seen from the angle of the theory of science the present situation reminds strongly of that at the transition from Ptolemaic to Copernican theory. We should keep that in mind. The irregularities of planetary orbits which Tycho assumed compare quite well to the present state of particle physics. In Tycho's time, at the level of macrophysics, bringing in harmony the Ptolemaic system with the visible movements of the planets would have required assuming a variety of planetary eccentrics and epicycles. In our time, at the level of microphysics, bringing in harmony the planetary model of the atom with all kinds of radiation and with measurements of electrodynamic states would require assuming so many different particles and states of particles that they would be less manageable than Tycho's eccentrics and epicycles had been. And even then the planetary model would still leave gravity unexplained. Both theories were but mental crutches serving to prolong the agony of a theory long disproved instead of letting it die in dignity.

IV

The planetary model of the atom is *too ambitious* and at the same time *too narrow*.

It is too ambitious because it presupposes that our knowledge of the structure of matter has reached the end of the road, and that we can and must therefore explain everything else in terms of that model. Having originated from a problem of philosophy, it still is mainly a philosophical theory. It describes what theoretical physicists believe to be, but explains nothing beyond itself. Thus it feigns knowledge where none exists, and by permitting explanations only within the scope of its model it obstructs the progress of science.

The planetary model of the atom is also too narrow because it will never be possible to explain the force of gravity in terms of that model. It cannot even explain the forces working within the atom itself because they too would have to be forces of attraction according to present theory. But no theory relying on particles alone was ever able to explain any force of attraction whatsoever. I even contend that no particle theory ever will be able to explain attractive forces through physical properties of the particles themselves unless, by equipping them with tentacles or with hooks and eyes or some other device of that kind, it makes them so complicated that they will be more in need of explanation than attraction itself. As longs as we retain the principle of locality defining 'physical' explanation as the description of the immediate action of one object on another, any explanation of an attracting force must at least consist in describing something acting *on* some object, and that cannot be the oject itself.

There is no reason to mourn the demise of the planetary model of the atom either. Though admittedly a beautiful visualization, it never contributed in any way to the progress of science. Niels Bohr's trying to bring together quantum theory and the periodic system of the elements was the only attempt ever at using it for explaining something beyond itself. If empirical evidence had confirmed his conjecture of electrons emitting and absorbing energy by jumping to smaller or larger orbits, then that would have been a genuine discovery. It would have explained the well-known fact that matter can emit and absorb energy by the hitherto unknown hypothesis that matter is composed of atoms consisting of a nucleus and one or more electrons orbiting it.

Without evidence however, and without quantum theory, Bohr's theory would first have had to explain how jumps of electrons might explain the emission and absorption of energy. Failing that, the planetary model cannot claim to be an explanation of anything. It is a beautiful model, possibly too beautiful, but it does not add in any way to our *physical* knowledge because it explains nothing.

Theoretical physicists will therefore have to get familiar with the idea that human knowledge is not nearing completion as many seem to think. We have no reason to believe that what contemporary science considers to be atoms or their components comes anywhere near to being the *ultimate* components of matter. Only sophomoric science can generate such notions tacitly implying that man is the measure of all things and approximately even of nature itself. In reality nature has set narrowest limits to our means of perception. Between those limits and what really would be the ultimate components of nature there may be dimensions of which we have no idea because they are too far beyond the scope of our perception. We should therefore be much more reticent already in the approach we use in our explanatory efforts. And we should consider, at least tentatively, that our possibilities of knowledge might still be lightyears away from what really are the *ultimate* limits of nature.

However, I repeat that giving up the planetary model need not imply giving up also the idea of a parallelism in the principles of order in the universe and on the subatomic level. The following chapters will show that we can even make that parallelism simpler and more convincing by adopting a new and stronger variety of the ether hypothesis. They will also show that falling back on ether theory is the only approach left open for ever finding an explanation of gravity.

CHAPTER 11:
EXPLAINING GRAVITATION

Gravity is most intriguing problem of theoretical physics; approach as in Popper versus Einstein I. Newton saw the problem; history of approaches: Ptolemaic theory would have explained motion of planets; Copernican theory explained only their observation; Newton described attractive force without explaining it. II. Einstein's four-dimensional space-time-continuum' confused gravitation and acceleration; also mathematics and physics; distinguishing them by Popper's definition of 'real' in his theory of 'Three Worlds'; ambiguity of term 'space'; Standard Model of Particle Physics cannot explain forces of attraction. III. Newton's laws of motion agree with empirical evidence; attempts at explaining gravity by 'push' imply returning to ether theory or giving up principle of locality; action at a distance would not be physical explanation; could only be last resort. IV. possibilities of explaining gravity by 'push' not exhausted; modern attempts assuming ether to be distinct from matter were ad hoc. V. Descartes' and Huygens' theories unknown to me when writing Albert Einstein oder: Der Irrtum eines Jahrhunderts, striking similarity; Michelson's experiments only refuted stationary ether, not Fresnel's theory; previous theories can be made stronger by assuming matter and ether to be different states of the same substance; hypothesis explaining many different phenomena; four conjectures: (a) ultimate components of ether so small that even indirectly observable only in large quantities; (b) ether is omnipresent: that in connection with motion implies elasticity, and that implies inertia; individual order breaks down but statistical order is created by uniformity of conditions (c) matter is not distinct from ether but itself a form of appearance of ether, a superimposition like waves in a river; (d) ether continuously decelerates in the direction of the centre of matter, backward impulse like in a traffic jam; explaining gravity in three steps: (1) no conflict with Newtonian theory; (2) stability of matter explained through principle of resonance; (3) deceleration of ether and stability of matter (resonance) combined result in acceleration of matter (gravity). VI. Other phenomena explained: Earth's rotation and vortex shape of hurricanes and galaxies; transportation of light and electricity, thunderstorms, magnetism, catalysers, effects of homeopathic medicines; possibility of crucial experiments demonstrated in 'Popper versus Einstein'; ITER fusion reactor decisive for my theory but not for rescuing the Standard Model of Particle Physics.

The problem of explaining gravity is not only the greatest of all the open problems of theoretical physics but also a striking demonstration of the limits of our physical knowledge. Newton and Faraday had clearly recognized it but found no solution. Einstein had tried to solve it in his general theory of relativity, but his approach was based on the formulae of his special theory which were self-contradicting as we saw in *Chapter* 8. Although that brings down also his general theory his approach nevertheless deserves being discussed because it was original and interesting. The presently prevailing theories relying on the standard model of particle physics not even try to explain gravity. The theoretical physicists living in our time seem simply to have given up.

For discussing the explanation of gravity it will be particularly important to keep in mind that the terms *descriptions* and *explanations* have different meanings in everyday language as well as in the language of science. I repeat therefore that as I use those terms, the words 'explanation' or 'to explain' will refer only to cases in which the empirical information contained in the *explicans* (the explaining words or sentences) goes beyond that contained in the *explicandum* (the state of affairs to be explained). The important feature of an explanation is that it adds something to the information we previously had. The words 'description' or 'to describe', on the other hand, I will use for statements not going beyond generalizations of information stated before or assumed to be known already[345].

My approach to explaining gravity still is the same as that shown in my *Popper versus Einstein* (1998). I will explain it in section *IV* of this chapter. For understanding the present situation we must begin by looking at the history of the problem.

---

345 These definitions refer to section *II, 1* of the *Introduction*, above.

I

Describing gravitation poses no problems since the times of Copernicus, Galileo, Kepler, and Newton. Newton's laws of motion not only enable us to predict the stage of the moon at any time of the month, or the positions of the planets during the year, or when some comet will next be visible again. They also enable us to fly aeroplanes, launch satellites, or hit distant targets. For all practical purposes their merits are unlimited and undisputed. Anyone content with getting reliable descriptions of physical effects is justified in considering gravitation as posing no problem at all.

Explaining gravity, on the other hand, has a long tradition as a problem of philosophy. Its history is well known but I must recall some aspects here as a background to the present situation.

Aristotle (384–322 b.C.) had believed the earth to be at the centre of the cosmos and her own centre to be the geometrical *locus* of all heavy things. As an attempt at explaining the force of gravity that begged the question because the *explicans* was as much in need of explanation as the *explicandum*. Aristotelian influence remained strong however, and cosmology long neglected gravitation although it raises great problems in that field. For all we know, the ancient heliocentric theory proposed by Aristarchos in the 3rd century b. C. never considered it.

The great merit of the theory of celestial spheres developed by Ptolemy in the 2nd Century A.D. was that it not only explained the motions of celestial bodies but also seemed to agree with the effects of gravity. If the invisible crystal spheres of Ptolemaic theory had really existed, then they would have explained the motions of the sun, the moon and the planets relative to other stars. To some extent they would even have explained even the irregularities in those motions. And the material quality of those spheres would also have explained also why celestial bodies do not fall on the earth. From the point of view of our need for explanations that would have been an improvement even on Newtonian theory which left gravity well described but unexplained. It would have satisfied the modern *principle of locality*, requiring that 'physical explanation' must consist in describing an immediate action of one physical body or entity upon another body or entity[346]. Ptolemaic theory was mistaken as we now believe to know, but seen from the angle of Popper's theory of science it was nevertheless a rationalistic theory. That may have been why it survived for more than a millennium although we now believe Aristarchos' much older heliocentric theory to have been nearer to the truth.

Ptolemaic theory only lost credibility when Copernicus and Kepler had shown a way of describing the motions of the planets in the much simpler terms of Kepler's laws of the ellipse. The drawback to their theory was that it left gravity unexplained. All it explained were the *observations* of the planets made from antiquity to the days of Tycho by assuming them to be visualizations of celestial bodies orbiting the sun, and by putting the earth one of those orbits. Its strength lay in its simplicity but its power of explanation ended there. Like Aristarchos' old heliocentric theory it did not say what caused the planets to stay on their elliptic orbits.

Galileo's laws of fall then described some of the effects of gravity in terms of acceleration[347], and based on that, Newton's *Principia* brought together gravitation and astronomy in one comprehensive theory. It described by identical principles the effects of gravity here on earth and the motion of the planets in space and became the most successful scientific theory of all times. In more than three centuries no one ever found it at fault.

Newton could not explain the cause of gravity however, and the passages quoted in *Chapter 1, I* show that he was fully aware of that[348]. On the one hand he saw that an attracting force of gravity could not be in-

346 For the relevance of that principle, see Popper, *Quantum Theory and the Schism in Physics* (1982), p. 19–21, 25, 26.

347 Galileo Galilei, *Discorsi i demostrazioni matematiche intorno a duenuove scienze attenenti alla mecanica & i movimenti locali* (1638), quoted from the German translation in *Galileo Galilei, Schriften, Briefe, Dokumente,* ed. by Anna Mudry, pp. 329ff., 375ff.

348 The quotations in *Chapter 1, I,* were from the *General Scholium* at the end of the *Principia* (Cajori, *Sir Isaac Newton's Mathematical Principles of Natural Philosophy and his System of the World,* vol. II, p. 546 f.) and from Newton's letter to Bentley dated Jan. 17, 1692 (Cajori, *Sir Isaac Newton's Mathematical Principles of Natural Philosophy and his System of the World,* vol. II, p. 633 f.).

herent in matter itself because that would mean giving up the principle of locality and admitting an occult force, acting at a distance. On the other hand he could find no evidence for an external cause of gravity. Seeing himself thus faced with the alternative of either giving up the notion of gravity being caused by attraction or admitting in physics miracles in the guise of action at a distance, he left the problem of causal explanation 'to the consideration of (his) readers'. Nevertheless, in *Definition V* of his *Principia* he put gravity into the category of 'centripetal forces' in the following words:

> '*A centripetal force is that by which bodies are drawn or impelled, or any way tend, towards a point as to a centre.*
>
> Of this sort is gravity, by which bodies tend to the centre of the earth; magnetism, by which iron tends to the loadstone; and that force, whatever it is, by which the planets are continually drawn aside from the rectilinear motions, which otherwise they would pursue, and made to revolve in curvilinear orbits.' [349] (Cajori's translation and italics).

Newton may have reasoned that only an attracting force would agree with the very simple and convincing laws of motion which he had found, and that it must therefore exist in nature. Or he may have stood under the influence of Aristotelian essentialism too strongly. There seems to be no way of knowing which interpretation is true. The wording of his *Definition V* seems to indicate that he believed in a 'natural tendency' of matter to move towards the centre of the earth. But both his letter to Bentley and the text quoted from his 'General Scholium' tell against that[350].

---

349 Cajori, *Sir Isaac Newton's Mathematical Principles of Natural Philosophy and his System of the World*, vol. I, p.2.

350 My own interpretation is that Newton was facing a dilemma which nobody could have solved in his time because it would have required physical knowledge obtained only centuries later. On the one hand he could not find an *explanation* of gravity, and on the other hand he could not give up his excellent theory *describing* the laws of motion. If he

The effect of Newton's great influence on posterity was, at any rate, that by his *Definition V*, he 'switched the principles of the universe from push to pull'[351]. The amazing achievements of his theory in all other fields silenced any criticism that might have been forthcoming. And there, with few exceptions to be mentioned below, the matter rested until Einstein addressed the problem again in his General Theory of Relativity (1916)[352].

II

Einstein's general theory was a truly original attempt at solving the problem of gravity. He had recognized it as a problem, he knew that his special theory had left it unresolved because it applied only to systems moving on straight lines and with constant velocity, and he had not noticed the mathematical self-contradiction in his special theory which we saw in the *Introduction* and in *Chapter 9, IV.*

In his general theory Einstein therefore set out to develop the formulae needed for applying the principles of his special theory to accelerated systems and to velocities on curved lines. Following an approach proposed by Minkowski[353] he assumed the three coordinates of space and the single coordinate of time to be 'mathematically equivalent' which inspired him to replace the concept of 'three-dimensional space' by that of a 'four-dimensional space-time-continuum'[354]. He also assumed gravitation and ac-

---

had mentioned that dilemma at the beginning of his *Principia* it would have impaired the success of his theory. By showing it only at the end of his book or in letters to friends he probably did the right thing by science.

351 I think the words between my quotation marks were Popper's but have not been able to locate them again in his books. For his interpretation of Newton's theory of gravitation see *Conjectures and Refutations*, p. 106, 107.

352 Einstein, *Die Grundlage der allgemeinen Relativitätstheorie* (The Foundation of the General Theory of Relativity), Annalen der Physik vol. 49 (1916), p. 669 f.

353 Hermann Minkowski, *Raum und Zeit* (1909).

354 Einstein, *Die Grundlage der allgemeinen Relativitätstheorie*, Annalen der Physik vol. 49 (1916), p. 773–777 (The Foundation of the General Theory of Relativity, Collected

celeration to be different aspects of similar situations[355]. Having noticed that seen from inside an accelerated system the effects of acceleration resemble those of gravitation[356], and that analytical geometry will depict the acceleration of bodies in curves whereas gravity exists in reality even where no acceleration is to be observed, he concluded that space itself must be curved. His next inference was that we must describe space in Gaussian 'curvilinear co-ordina tes'[357]. He further assumed that gravitation would influence the 'metrical properties of space' and hoped that integrating in his concept of a 'four-dimensional continuum' the parameters of time would then enable him to develop a formula expressing the effects of gravity[358].

(1) Seen from the angle of modern theory of science it is evident that besides overlooking the self-contradiction in his fundamental premise of the constancy of the speed of light, Einstein also confused mathematics and physics. He knew that physics must stand up to the test of reality, but he seems not to have realised that mathematics and geometry are non-empirical because they are part of man-made human language where we ourselves *create* the meaning of what we say, and where we therefore bear responsibility for that meaning. That was why he went wrong already when he assumed the three coordinates of space and the coordinate of time to be 'mathematically equivalent'.

Papers vol. 6, p. 151–154).

355   Einstein, *Die Grundlage der allgemeinen Relativitätstheorie*, p. 773 (The Foundation of the General Theory of Relativity Collected Papers vol. 6, p. 150).

356   Einstein, *Die Grundlage der allgemeinen Relativitätstheorie*, Annalen der Physik vol. 49 (1916), p. 772, 773, 779 (The Foundation of the General Theory of Relativity Collected Papers vol. 6, p. 150, 151, 156); *Über die spezielle und die allgemeine Relativitätstheorie* (On the Special and General Theory of Relativity, 1917), p. 49.

357   Einstein, *Grundzüge der Relativitätstheorie*, p. 63, 64. (The Meaning of Relativity, p. 65, 66).

358   Einstein, *Die Grundlage der allgemeinen Relativitätstheorie* p. 779, 819 (The Foundation of the General Theory of Relativity, Collected Papers vol. 6, p. 156, 196); *Über die spezielle und die allgemeine Relativitätstheorie* (On the Special and General Theory of Relativity), p. 49–51; *Grundzüge der Relativitätstheorie*, p. 57–65 (The Meaning of Relativity, p. 63–66).

It may have been his essentialist approach that led him astray at this point. That interpretation is conjectural of course, but if someone had convinced him that the 'essence' or the 'true meaning' of coordinates was in their mathematical equivalence, then that might explain why he mixed together the coordinates of space and time in his concept of 'curved space'. It would also show the dangers of that method consisting in declaring 'essential' some features of a concept and 'accidental' some others and then drawing from that evaluation inferences that neglect the differences between them.

Einstein's assumption that the rules of mathematics apply to the three coordinates of space as well as to the single coordinate of time is correct only partly. It is correct in so far as the *quantities* of the units are concerned in which we grade those coordinates. Those quantities will be expressed in numbers, and we can apply the rules of arithmetic to those numbers. This mathematical equivalence does not extend to algebra or to analytical geometry however. It neither includes the meaning of the units themselves nor that of the coordinates on which we depict them. We cannot express distance in seconds, or time in metres, nor add seconds to yards, or subtract miles from hours. The units we use for measuring distance and time stand for different sets of facts, and so do the coordinates on which we depict them. Mixing them together in a hazy concept of a 'four-dimensional space-time-continuum' is putting the cart before the horse because the coordinates themselves as well as the units we insert in them are not part of physical reality but of man-made language. We can draw coordinates on paper with a pencil. But we cannot discover them anywhere in space because they are part of the language of physics which we use for speaking *about* physical reality and for making calculable its aspects. Space itself is part of reality of course, and we may wonder whether space in that realistic sense is empty or contains dark matter, or black holes, or ether, or whatsoever else. But we must distinguish space in this physical sense from the *word* 'space' composed of five letters. That word is not part of physical reality but of language. In science it is important to distinguish between the objects of which we speak and the language we use for speaking of them, particularly if language itself is the object of which we speak. For making clear that distinction and for avoiding ambiguities we should observe Alfred Tarski's rule that

> 'whenever, in a sentence, we wish to say something about a certain thing, we have to use, in this sentence, not the thing itself, but its name, or designation'[359].

When speaking in meta-language of language itself in its role of object language, the best way of observing Tarski's postulate is by putting in quotation signs or writing in italics the terms of object language which we are discussing. Without distinguishing between those different levels of language every word of common language would have at least two meanings, itself and its designation.

In that sense the word 'space', composed of five letters, is not part of physical reality but of language. It can stand there as a symbol for designating various objects of which two are relevant in our context. It can stand for the infinite space of our universe which is an element of physical reality. But it can also be used as a term of geometry standing for an abstract situation in which three straight lines with angles of 90° between each two of them can be drawn through one point. In that sense the word 'space' stands for an element of geometry, and therefore of the man-made language of mathematics.

It may seem unfortunate that a term so important as 'space' should have more than one meaning but anyone desiring to express himself clearly can avoid ambiguity by using terms like 'real space' and 'geometrical space', or some other clarifying epithet. At any rate there is a difference between those meanings. And contrary to Einstein's view, 'space' in the geometrical sense does not normally stand for anything having physical properties of its own[360]. We could alter that of course, and decide to let

---

359   Alfred Tarski, *Introduction to Logic* (1965), p. 58–60.

360   For Einstein's view of space having 'physical properties' of its own see his *Äther und Relativitätstheorie* (Ether and the Theory of Relativity), Collected Papers of Albert Einstein, vol.7, p. 160. I quoted the relevant passage at the beginning of *Chapter 9, I.*

'geometrical space' stand for something having physical properties. That would hardly be convenient however because we would then have to find a new name for an *abstract* situation in which three straight lines with angles of 90° between each of them can be drawn through one point *without* having physical properties.

(2) Einstein's approach to solving the problem of gravity in his general theory is invalidated thus not only by the self-contradiction in his fundamental premise but also by the muddle in his concepts. That may have been why it never convinced many.

At the end of the 20th Century at any rate some of the most famous physicists of that time, Richard Feynman, Leon Lederman and Stephen Hawking among them, and none of them notorious for being critical of the theory of relativity, nevertheless agreed that gravity still was unexplained. Feynman wrote in 1985 that in his time there did not even exist a reasonable *attempt* at explaining it[361]. And Lederman and Hawking both believed the explanation must lie in some particle. They even gave it names like 'graviton' or 'gravitino' but no scientist ever was able to find it[362]. It has nothing to do with the Higgs boson recently said to have been discovered[363]. The scientists at the CERN make no efforts to search for it because they know not even what kind of physical properties they should be looking for. Some theoretical physicists argue that physical theory can neglect gravity because it is but a weak force compared to other forces working within the atom. They fail to see however that on the strength of that argument we would have to neglect all those other forces first because an atom is so small compared to a heavy stone.

Explaining gravitation thus remains the greatest unresolved problem of theoretical physics in our time. Newton had already faced the alternative of either giving up the notion that gravity is a force of attraction or

---

361   Richard P. Feynman, *QED – The Strange Theory of Light and Matter*, German translation p.171.

362   Leon Lederman, *The God Particle* (1993), German translation p.9, 528, 530; Stephen Hawking, *A Brief History of Time* (1988), p. 70, 157.

363   According to CERN's press release of March 14, 2013, the discovery was made in 2012.

admitting action at a distance. That alterntive still is open, and its full significance shows in the fact that gravity is not the only attracting force of nature which particle physics cannot explain. It can *describe* any ostensively attracting force of nature in terms of the standard model. But when it comes to *explaining* such forces, then that model theory is faced with the same problem that Newton had been facing, and is equally at loss. It cannot say by what mechanism the strong interaction keeps the nucleus together. It cannot even say what causes the attracting or repulsive forces of a simple loadstone. All it can do is to assume the existence of 'magnetic fields' or 'field lines', or of some other kind of 'fields of forces'. Theories of that type may be useful descriptions for practical purposes but they are no explanations because they only replace one term of uncertain meaning by an equally uncertain one without telling us anything new about reality.

The Standard Model of Particle Physics thus is faced with the problem of all model theories in science. Instead of explaining the known by the unknown it attempts to explain the unknown by the known. That is why it has a tendency to hamper new discoveries because it can tolerate explanations only within the scope of its model whereas anything going beyond that would refute the model itself.

III

Any attempt at finding a physical explanation of gravity must bear in mind that for all we know Newton's laws of motion are true descriptions of reality, and that matter will *in fact* accelerate towards the centre of the Earth as long as no other body prevents it from doing so. Denying that would be neglecting all available empirical evidence. But acceleration can be caused by *push* or by *pull*, and if we could find a way of explaining it by push, then it might still be possible to reconcile Newtonian theory with the principle of locality .

Many have investigated that approach too of course, Newton himself being among the first[364]. They all assumed that a pushing effect causing an acceleration of matter cannot come from matter itself but must come from some other substance surrounding it or flowing through it. That would mean going back to some variety of the ether hypothesis which was given up in favour of Einstein's theories in early 20th Century. The issue at stake thus lies in either giving up locality or returning to ether theory.

The problem is urgent because the principle of locality has lately been called into question[365]. My own view is that giving it up and admitting action at a distance in physics would be equivalent to the end of physics as a science[366]. Popper argued that accepting the possibility of simultaneous physical effects without physical transmitter would be abandoning the principle of causality itself[367]. I agree with him that inferring from this that action at a distance *cannot* exist would be erroneous. If it did exist however, then we could explain by action at a distance any physical effect whatsoever. There would not be much point any more in searching for physical explanations and in making discoveries by introducing hypotheses with new empirical content. The principle of locality is the motor of all progress in physical knowledge. We should leave no stone unturned before giving it up[368]. That was Faraday's view too by the way, and he clearly saw the conflict between the principle of locality and Newtonian physics assuming gravity to be a force of attraction. I already quoted a short passage

---

364 See Newton's letter to Boyle of Feb. 28, 1678, quoted by Florian Cajori in Appendix to *Sir Isaac Newton's Mathematical Principles of Natural Philosophy and his System of the World*, vol. II, p. 675. - Some of the later efforts and their discussion have recently been collected and published in the volume *Pushing Gracvity – New Perspectives on Le Sage's Theory of Gravitation*, edited by Matthew R. Edwards, 2002. I know of only one new attempt made after that Duncan W. Shaw, *The cause of gravity – a concept*, Physics Essays 25, 1 (2012).

365 Aspect, Grangier, Roger, *Experimental Realization of Einstein-Podolsky-Rosen-Bohm Gedankenexperiment: A New Violation of Bell's Inequalities*, Phys.Rev.Lett. 49 (1982), p. 91.

366 *Popper versus Einstein*, Chapter 8, p. 168–176.

367 Popper, *Quantum Mechanics and the Schism in Physics*, p. viii, 25, 26.

368 See my *Popper versus Einstein*, p. 175; also, my *The Oscillation Project with Emulsion-Tracking Apparatus (OPERA) experiment: An argument for superluminal velocities?* Physics Essays, vol. 25 (2012), p. 401.

from his letter to Reverend Jones but it deserves a longer quotation at this point. Faraday wrote there:

> 'The cases of *action at a distance* are becoming in a physical point of view, daily more important. Sound, light, electricity, magnetism, gravitation, present them as a series. The nature of sound, and its dependence on a medium, we think we understand pretty well. The nature of light as dependent on a medium is now very largely accepted. The presence of a medium in the phenomena of electricity and magnetism becomes more and more probable daily. We employ ourselves, and I think rightly, in endeavouring to *elucidate the physical exercise* of these forces, or their sets of antecedents and consequents, and surely no one can find fault with the labours which eminent men have entered upon in respect of light, or into which they may enter as regards electricity and magnetism. *Then what is there about gravitation that should exclude it from consideration also?* Newton did not shut out the *physical view*, but had evidently thought deeply of it, and if he thought of it, why should not we, in these advanced days, do so too? *Yet how can we do so if the present definition of the force, as I understand it, is allowed to remain undisturbed?* Or how are its inconsistencies or deficiencies as a *description* of the force to be made manifest, except by such questions and observations as those made by me, and referred to in the last pages of your paper? I believe we ought to search out any deficiency or inconsistency in the sense conveyed by the received form of words, that we may increase our real knowledge, striking out or limiting what is vague. I believe that men of science will be glad to do so, and will even, as regards gravity, amend its *description*, if they see it is wrong. You have, I think, done so to a large extent in your manuscript, and I trust (and know) that others have done also. That I may be largely wrong I am free to admit – who can be right altogether in physical science, *which is essentially progressive and corrective?* Still, if in our advance we find that a view hitherto accepted is not sufficient for the coming development, we ought, I think (even though we risk something on our own part), to run before and rise up difficulties, that we may learn how to solve them truly. To leave them untouched, hanging as

dead weights upon our thoughts, or to respect or to preserve their existence whilst they interfere with the truth of physical action, is *to rest content with darkness and to worship an idol.*[369] (My italics).

Those are heart-warming words, revealing the spirit of a true scientist. Faraday did not even pretend he had solutions for everything. Though stating without reservations his discontent with the theory prevailing in his time, he freely admitted being unable to present a better one. Yet he would not be tempted into accepting an attracting force because that would have meant 'to rest content with darkness and to worship an idol'. His firm belief in the necessity of the principle of locality was stronger than his desire to present a ready-made solution to the problem of gravity.

That was in a distant past now. In our days the principle of methodological nominalism can provide us with an even stronger argument for agreeing with Faraday. The problem of explaining gravity is not only in the question of whether or not there is action at a distance but also in distinguishing physical explanations from other kinds of explanation. The term 'explanation' itself can stand for various types, for instance for medical explanations, or psychological explanations, or historical explanations. If we give up locality in physics, then that would give the term 'physical explanation' a meaning that includes physical effects without a physical transmitter. It would apply to parapsychological explanations assuming telepathic effects and to theistic explanations assuming God as the agent. That would seriously impair the clarity of the concept 'physical explanation' and of the language of science in general. In an empirical science like physics, confessing our own ignorance by admitting, as Newton and Faraday did, that we are faced with a situation for which we can find no physical explanation is far more straightforward than calling 'physical' an explanation that does not add to our understanding of physics.

---

369 Quoted from Peter Day, *The Philosopher's Tree, A Selection of Michael Faraday's Writings* (1999), p.104f.

The upshot of all this is that we must fall back on ether theory if we ever want to find an explanation of gravity deserving the epithet 'physical'. Can it be of any avail however? Where Huygens, Newton, Faraday and so many others failed, the situation seems hopeless from the outset.

In spite of those numerous failures I contend that the possibilities of explaining the effects of gravity 'by push' have not yet been exhausted. All previous attempts started from the assumption that a pushing effect causing the acceleration of matter cannot come from matter itself but must come from some other substance surrounding it or flowing through it. That applies to the approaches taken by Descartes, Huygens and Newton[370] and to all the more modern theories of the Le Sage-type recently collected in the volume 'Pushing Gravity'[371]. It also applies to Duncan Shaw's latest proposition to explain gravity through a deceleration of ether caused by a 'ram force' as it traverses matter[372]. All these theories assume that ether is something distinct from matter, and that it somehow pushes matter in the direction of the centre of gravity.

We need not discuss those theories in detail here. Some of them have much in common with the explanation I am going to propose. My chief objection against all of them is that they *only* try to explain the well-known effects of gravity such as acceleration in the direction of the centre of matter, or the fact that no shield or filter can interrupt it, but nothing beyond them. That stamps them with the stigma of being mere *ad hoc*-hypotheses. It is legitimate to equip ether hypothetically with all those physical properties such as texture, or elasticity, or velocity or whatever

---

370 Descartes *Traité de la Lumière* (1664); Huygens, *Traité de la Lumière* (1690); for Newton see Cajori, *Sir Isaac Newton's Mathematical Principles of Natural Philosophy and his System of the World*, vol. II, p. 675, where he quotes from Newton's letter to Boyle, dated 28 February, 1678.

371 Pushing Gravity, New Perspectives on Le Sage's Theory of Gravitation, edited by Matthew R. Edwards (2002).

372 Duncan Shaw, *The cause of gravity – a concept*, Physics Essays 25, 1 (2012), p. 66ff.

else we need for explaining gravity. There is no other way of finding out whether ether exists than by assuming it to have physical properties and then searching for their traces. If a theory is plausible, then it will be the task of experimental physics to test it. We might have to fall back on one of the theories just mentioned therefore, but I still believe we can do better than that. We can conceive a variety of the ether hypothesis that will explain more than the theories just mentioned, and we can design more experiments for testing its truth empirically.

I first stated my proposition for explaining gravity in *Popper versus Einstein* (1998)[373]. Before giving account of it here I must confess that at the time of writing that book I had read about the works of Descartes and Huygens but never read them in the original. Only a critic of my book drew my attention to Descartes by pointing at a likeness between his vortex theory and the one I had proposed. When writing *Albert Einstein oder Der Irrtum eines Jahrhunderts* (2009), I had read Descartes' *Traité de la Lumière* (1664) but had still not read Huygens' treatise of 1690 bearing the same title.

I deeply regret that now because reading Huygens' *Traité de la Lumière* would have spared me many difficulties. I found it on the Internet only while working on this book, and was surprised and grateful to see that my own explanation of gravity agrees with his in all respects but one. I will mention the similarities as we go along but the difference between our views is more important.

Huygens discussed only the theory of light in his treatise. He mentioned gravity in passing in Chapter III, but he did not mention the problem of explaining it. That may be the reason why our approaches differ in one important aspect. Whereas Huygens assumed ether to be distinct from matter, I think we must replace that by the hypothesis that *ether and matter are different states of the same substance*. We will see that this opens interesting possibilities of explaining not only gravity but also several other physical phenomena.

---

373 *Popper versus Einstein*, Chapter 7 (p. 141–155).

(*1*) In my theory ether is not distinct from matter and surrounding it or passing through it. Instead, I assume *matter itself to be composed of ether,* and its manifestations to be a state of ether, a superimposition as it were. The relation between ether and matter would then be somewhat like that between water and its waves, or between air and the sound travelling through it. What we perceive as the substance and the stability of matter would then in fact be a uniformity in the state of that superimposition, lasting only as long as the conditions do that create it.

To the best of my knowledge no one before me proposed this hypothesis which I will state in outline here. I believe it to be the strongest variety of the ether hypothesis. It is well in line with empirical evidence available so far, and it will explain more physical phenomena than any previous variety did. Through the physical properties of ether it can explain not only the phenomena of light but also those of matter itself. That makes it simpler than all previous ether hypotheses and also simpler than all attempts at explaining matter as consisting of atoms or their components.

Its simplicity lies in the fact that it explains a large variety of phenomena from the single principle of elastic ether. On the other hand it requires some power of imagination. That should be not held against it however because the situation at the transition from Ptolemaic theory to Copernican theory was not different. And besides, the explanation of gravity would hardly have remained undiscovered for so long if it had been easy to find. If we consider returning to ether theory we should therefore address ourselves to this new variety primarily. In *Popper versus Einstein* I proposed some experiments that would refute it if it should be mistaken[374]. I will mention them further down.

In briefest outline I assume ether to consist of highly elastic particles in constant motion and filling what we believe to be the universe. The ether particles themselves are so infinitesimally small that we must neglect their individual existence, far smaller than anything the standard model of particle physics would permit. Though ultimately composed

---

374 *Chapter 10* of my *Popper versus Einstein* (1998).

of particles, ether nevertheless *behaves* like a fluid therefore[375]. We may visualize the universe as a cloud then, or as an avalanche in which, regarding proportions, the relation of individual ether particles to *material* bodies here on earth might resemble that of the earth to celestial galaxies. In that infinite ocean of moving ether, matter is like a wave in a river or like a cloud in the sky, its substance changing as the water molecules flow through it, but the wave itself retaining its outward appearance as long as the conditions persist that shape it. The physical properties of ether, in particular its elasticity and the principle of resonance, will then explain the stability of matter.

(2) Starting from that notion we can replace the standard model of particle physics by a different theory of matter. It may seem strange at first glance but it will open an approach to explaining not only gravity itself but also several other physical phenomena hitherto unexplained without giving up any part of Newtonian physics.

Fresnel stated his hypothesis of moving ether in 1818. Adding to it only a few simple conjectures will make it much stronger than he himself had stated it. Provided it stands up to empirical criticism, the new theory will not only improve our understanding of nature but also raise new questions. It will open new fields for scientific imagination and will enhance our consciousness of our own ignorance. Before coming to that, we must see the conjectures. I will explain them as we go along.

Conjecture (a) is that the ultimate components of ether, possibly consisting of various different types or shapes, are so small that all we can observe of them *even indirectly* are large quantities of them[376].

---

375 There is no contradiction in assuming ether to be elastic, yet behave like a fluid. Water is a fluid but it is also elastic. Otherwise, the effects of the great tsunami in 2004 would be inexplicable. We saw that in *Chapter 9, VI, 1,* above.

376 Huygens and Newton's both conjectured that ether must consist of several different kinds of particles. For Huygens, see his *Traité de la Lumière*, p.14-18; for Newton, see his letter to Boyle of Feb. 28, 1678 quoted by Cajori in the Appendix to *Sir Isaac Newton's Mathematical Principles of Natural Philosophy and his System of the World*, vol. II, p. 675.

That does not exclude the possibility of their being particles but it means that we must return to wave-theory nevertheless [377].We must bear in mind how little we know, and that *infinity has two directions*. It goes not only in the direction of outer space but also in the direction of microphysics. Time, too, need not be measured on terrestrial scales. We know that compared to the extension of the universe we are no more than a speck of dust and hardly that. But we sometimes forget that compared to the ultimate components of nature we may be larger than the whole of our galaxy is compared to us.

Our means of perception are extremely limited. We have extended them by inventing telescopes and microscopes and by indirectly making visible some invisible effects, and we have found other means of indirectly gathering information about this world. Yet although we have been quite clever we nevertheless have no reason to believe that the phenomena which we poor human beings can only observe come anywhere near to being the limits of nature itself, nor that the periods of time covered by human observations are relevant for understanding the ultimate laws of nature. We know that all matter will decay but we must learn to accept that even what we perceive as the stability of matter in the history of our galaxy may still be a dynamic process in the evolution of the universe.

If we return to the ether hypothesis, then we must accept that the ultimate components of ether may be much smaller than anything the standard model of particle physics would ever permit. They may be so infinitesimally small that all we can observe of them *even indirectly by their effects* are not particles but only large quantities of them in the shape of clouds, or clusters, or vortices. What we observe in cloud chambers and interpret as particles or their traces would then not be individual particles but large quantities of far smaller components of ether. We may not take

---

377 That conjecture comes fairly close to Heisenberg's view, as explained in *Das Naturbild der Physik,* p. 124 (see the quotation in the footnote at the beginning of *Chapter 10*). The important difference however is that Heisenberg assumed a 'dualism' of waves and particles (which is an explanation at the level of language) whereas I assume the ultimate components of ether to be so small that all we can observe even indirectly are statistical results of large quantities of them (which is a conjecture about physical reality).

the physical properties of those clusters for physical properties of the ultimate components themselves therefore because that would be a serious fallacy of method[378]. Instead, we must accept that although we assume ether to be composed of particles, we nevertheless can observe *only* the clusters and their reactions. Even with the best of our instruments we may still be in the situation of one who can observe the waves of a river but not the molecules of water flowing through them.

Conjecture (b) is that ether is omnipresent, elastic and in permanent motion. We do not know the origin of that motion. A collision between two ether streams flowing in opposite directions would be plausible but it would not get us much further. I will discuss some methodological aspects of that situation in the following chapter. At any rate omnipresence and motion of ether entail a number of physical implications needing explanation[379]. I mention only some of them here, leaving open to what extent they are implications in the strict sense or conjectural themselves.

If ether is omnipresent, then we must assume it to be elastic. Without elasticity a universe filled with ether would have to be one static block in which no impulse would be possible. Elasticity on the other hand, as we observe it for instance in the steel balls of *Huygens' cradle*[380], requires a certain flexibility of material and a tendency to resume its previous state after receiving an impulse.

Flexibility and resuming its previous state both imply motion, and motion is a process, therefore implying time. Elastic ether, whatever its ultimate components may be, will not resume its previous state instanta-

---

378 The best way of explaining that is by an example. Measurements on a motorway may yield an average speed of the cars of 50 even if not a single car rolled at that speed. Similarly, assuming elasticity to be a function of the extent of deformation caused by an impulse and the time taken for resuming its previous state, a mixture of materials may have an elasticity of x even if none of its components have exactly that elasticity but half of them have x + 10 and the other half have x - 10. For the methodological aspect see also Popper, *Quantum Theory and the Schism in Physics*, p. 52 f.; v. Mettenheim, *Popper versus Einstein*, p. 126–128.

379 For those implications, see also *Popper versus Einstein*, p. 119 f.

380 For 'Huygens' Cradle', see *Figure 4* in *Chapter 9, VI, 1*.

neously after receiving an impulse therefore but only some fractions of a second later, even if they be infinitesimally small. Elasticity thus implies also a delay of reaction to an impulse. Considering that effect on a larger scale its result is *inertia*, and it explains also the phenomenon of *entropy*.

As a result of that conjecture and its implications, which are elasticity and motion, any order that might once have existed in the ether filling the universe must have broken down almost from the outset[381]. In physics we assume similar conditions to cause similar reactions because no physical explanation would be possible without that assumption. Adjacent components of ether receiving impulses from outside will therefore tend to parallel reactions, which is the principle of *resonance*. Due to their inertia however, that tendency will decrease with increasing distance from the impulse. And due to the omnipresence of ether its components will sooner or later meet with impulses from others, acting in different directions and creating *torsional effects* at the point of every individual collision that is not precisely head-on. At the level of the ultimate components of ether that process must eventually result in chaos. But at the level of the clouds or clusters of ether, which are all we can observe even indirectly (conjecture *a*), some kind of *statistical* order may be created by the uniformity of situations and reactions.

Conjecture (c) is that *matter* is not distinct from ether but is itself a form of appearance of ether, a superimposition as for instance the waves in a river are superimpositions on water, or sound is a superimposition on air[382].

---

381 'Huygens' Cradle' (see *Figure 4* in *Chapter 9, VI, 1*) demonstrates that effect. It usually consists of five steel balls standing in direct contact with one another, each of them suspended on two strings to keep them in line. Letting ball No. 1 bounce against ball No. 2 will at first bounce off only ball No. 5 while balls Nos. 2–4 retain their position. If the process goes on however, that order will gradually break down, and eventually all balls will be in a state of unrest. The cause of that disorder is not in lack of precision of the instrument, but in the fact that their elastic responses to each impulse are not instantaneous but slightly delayed. It thus is an experimental demonstration of entropy.

382 That is the most important difference between my theory and the one Duncan Shaw recently proposed (Physics Essays 25 [2012], 1). Shaw was not aware of the theory I had stated in *Popper versus Einstein* (1998) when he wrote, but we both agree that gravity must be explained by 'push', therefore by deceleration. We also both assume ether to

I will explain it in more detail as we go along.

Conjecture (d) is that ether *decelerates* in the direction of the centre of matter.

The deceleration of ether may be implied already in conjecture (*b*) assuming ether components to be elastic and in permanent motion. Given that, it is reasonable to assume that their motion will stop or almost stop in head-on collisions with other ether components. Their stopping or almost stopping will cause a compression in the sequence of other components approaching the point of collision from the same direction and standing in direct contact with one another. And that impulse, caused by compression, will result in their gradual deceleration as we observe it in 'Huygens' Cradle'[383].

(*3*) Based on those conjectures, which are reasonable from the point of view of ether theory, we can explain gravity in three steps.

Step (1) is to show that given moving ether there could be a way of explaining an acceleration of matter through the deceleration of ether without giving up Newton's laws. That is its *only* purpose. It is not part of the explanation of gravity itself.

Let a railway carriage roll along at constant speed and a modern suitcase with rollers underneath stand in its corridor. In an emergency braking of the train that suitcase will roll forward until it crashes into the front wall of the carriage. If the emergency braking causes a deceleration of neg-

---

flow through matter. However, whereas Shaw takes ether to be something distinct from matter, I believe matter to be a form of appearance of ether. In his theory the relation of matter and ether is like that between a tube and the water flowing through it, and in mine it is like that between water flowing in a river and the *shape* of the waves, or cataracts, it forms. The difference is important because Shaw's theory, assuming a ram pressure exerted by ether, and greater in ether flowing into cosmic bodies than in ether that is being expelled into space, may be able to explain gravitation but it will not explain any of the other effects mentioned in the text below. That is why I think that seen from the angle of the theory of science, it is a mere *ad-hoc*-hypothesis.

383 See *Figure 4* in *Chapter 9, VI.*

ative g, then, according to Newton's first law of motion, the acceleration of the suitcase relative to the carriage will ideally be positive g[384].

The acceleration which we observe in gravitational effects of matter might thus originate in the deceleration of the ether of which it consists. This shows that conjectures (*b*) and (*d*) need not lead to any conflict with Newtonian physics.

Step (2) requires more imagination. It is also not to explain the force of gravity but only to show that the elasticity of ether and the principle of resonance can together explain the *stability* of matter.

If elastic ether is in permanent motion and continuously decelerating in the direction of the centre of matter (conjectures *b, d*), then each *material* particle, itself consisting of a much larger number of compressed ether particles (conjecture *d*), will constitute its own tiny centre of deceleration. And if ether is elastic, then equal impulses will cause equal reactions (conjecture *b*). Those resonances will permit intermediary states between resounding oscillations only within the limits set by the elasticity of their material. The regularity of those reactions in the large clusters of compressed ether particles shaping one material particle (conjecture *a*) is what we perceive as the stability of matter. Even the stability of matter is no static state in this view but a dynamic *process*.

Step (3) is to show how the deceleration of ether will cause an acceleration of matter.

We consider one single *material* particle under the influence of gravitation. We cannot call it an atom or a nucleus because we no longer believe in their existence after reading *Chapter 10*. The material particle must now be something far smaller than the nucleus of atomic theory would have been. Yet it consists of a very large number of ether particles. In fact, it *is* ether itself, but in a state different from that of the ether surrounding it. It might be like a small vortex in a river, remaining stable for some moments but soon to disintegrate as the river flows on. We assume that material

---

384 The word 'ideally' is to indicate that the argument neglects the friction of the rollers.

particle to be drifting along in the huge river of ether filling all space and to enter the gravitational field of the Earth. The flux of ether within that material particle must now decelerate, and its deceleration will increase as it approaches the centre of gravitation (conjecture *d*). But what will happen to the material particle itself?

It is not infinitely small but occupies a physically extended part of space. In relation to the ultimate components of ether or to the unlimited possibilities of smallness which we have to take into account (conjecture *a*), its extension is not negligible. Since all material particles consist of ether (conjecture *c*), even a single material particle can be gigantic in relation to the ether particles composing it. And we assume a *continuous* deceleration of ether towards the centre of matter (conjecture *d*).

The front of our material particle will therefore meet with stronger deceleration of ether than its rear which is lagging behind in relation to the size of the particle. This means that the front will receive a backward impulse but that the resonance between the ether particles *within* the material particle, caused by their uniform elasticity, will be stronger than the *external* impulse caused by the deceleration of the surrounding ether. It will therefore hinder a major change in the diameter of the material particle (*Step 2*). Thus its stability, caused by the resonance of ether within the material particle, will hinder it from fully giving way to the external impulse of deceleration caused by the surrounding ether. The material particle will therefore have a tendency to retain its diameter, which means that its shell and its components, front wall and back wall, must all have a tendency to assume the same velocity. That implies that the material particle *cannot decelerate to the same extent* as the immaterial ether surrounding it.

This principle seems to be well in line with empirical evidence. In fact, I think it consists only in transferring to ether what we already know about the air and other gasses here on earth. The speed at which sound travels is correlated to the density of the medium in which it travels. It travels faster in water than in air; and it travels faster at high temperatures than at low temperatures. The general principle underlying that seems to be that the denser a medium, the faster will its undulations travel through it. It is natural then to assume that density to influence also the resistance against a change of motion, be it acceleration or deceleration.

If the material particle cannot decelerate to the same extent as ether surrounding it, *then it must 'overtake' ether* as it were. The situation will be somewhat like a rear-end accident of motor cars. If the first car decelerates stronger than the one following it, then there is an acceleration of the second relative to the first, and if that process continues, then it will end in a crash.

(4) Only one principle will thus explain gravity, and entropy, and the inertia of matter. Ether in the compressed, decelerated and resonating state of matter will show more resistance against *further* deceleration than ether in its un-compressed, un-decelerated and un-resonating immaterial state[385]. Or looking at it from the other side, ether in its un-compressed and un-resonating immaterial state will be decelerated more strongly by resistance from obstacles standing in its way of motion than ether in its compressed and resonating state of matter.

The deceleration of ether will thus cause a relative acceleration of matter which will persist until other matter interrupts it according to Newton's first law of motion, or until there is an electrical discharge in which the ether oscillations will re-arrange themselves in a new harmonious resonance.

V

I already explained all this in *Popper versus Einstein*, some of it in more detail. Yet it remains but an approach to a theory that may one day lead to a better understanding of nature. The hypothesis of decelerating, elas-

---

385 In *Popper versus Einstein*, p.149, I had assumed that matter will not decelerate at all. For explaining gravitation it is sufficient however to assume that there is a difference between the deceleration of ether and that of matter.

tic and resonating ether deserves further consideration. It seems to be the only way left open for explaining the force of gravity in accordance with Newton's laws of motion, and for explaining by the same principle why it is impossible to interrupt that force by any shield or filter whatsoever.

The hypothesis of decelerating elastic ether opens approaches to explaining a number of other effects remaining unexplained at present. The transportation of light is one of them, and the underlying principle applies also to explaining electricity or thunderstorms. Catalysers and the effects of homeopathic medicines will also never be explicable in particle theory, but a wave-theory might explain them through the influence of resonance caused by adjacent materials. That approach solves also the problem of longitudinal and transversal waves because an elastic ether particle would respond to an impulse not only by reducing its diameter in the direction of the impulse but also by expanding its diameter sideways, though with less intensity[386]. I could name more examples but the rotation of the earth and the vortex shape of hurricanes and galaxies are the most striking. I will discuss them in the following chapter.

In *Popper versus Einstein* I proposed two experiments for putting to the test this hypothesis[387]. Constructing the fusion reactor ITER in France

---

386   That solves also the problem of infinite divisibility of energy mentioned in the footnote to *Chapter 8, III, 2,d*. If we explain the transport of energy by the elasticity of ether working in different directions, then no reason is conceivable why an increase or decrease of energy should not be continuous.

387   *Popper versus Einstein*, p. 189–195. - One of the experiments would rely on the principle, which Rœmer applied in his discovery of the finite speed of light. If light decelerates when approaching matter, as I assume it to, then the velocity of light coming from a distant star should be higher if it reaches the earth *before* traversing the gravitational field of the sun than *after* traversing it. Testing that would be possible by measuring the differences in the velocity of that light not in intervals of six months but of three months, therefore on half the diameter of the earth's orbit (see *Popper versus Einstein*, p. 190). – The other experiment would rely on terrestrial measurements of the speed of light made at different altitudes (p. 159f., 191f.). According to the theory proposed here we would expect lower velocities nearer to the centre of gravitation, therefore at lower altitudes. Measurements made by Michelson seem to confirm that (see Jaffe, *Michelson and the Speed of Light*, p. 164f.), but they were made under different aspects and may therefore not yet be sufficiently reliable.

had not begun then. When that reactor goes into action, if it ever does, then that will be the most important experiment for testing my theory.

Earlier versions of the ITER-website correctly pointed out that we cannot simulate here on earth the gravitational conditions existing on the sun. That implies however that we could not simulate in a fusion reactor the conditions of *pressure* existing on the sun or in a nuclear explosion because they would blow up the reactor. That difference of conditions might seem unimportant from the point of view of a theory assuming matter to consist of incompressible particles, but assuming matter to consist of *compressible* vortices of *elastic* ether, I think the difference of conditions of pressure in the sun or in a nuclear explosion compared to those in a fusion reactor will be decisive. That is why I do not expect the ITER-experiment to succeed, no matter how much money it will cost.

However, my chief objection against the standard model of particle physics is independent of the outcome of that experiment or any other. A model theory of matter that cannot explain in terms of its model the most obvious of all the forces of nature, which is gravity, cannot claim to be a *true* description of nature, no matter how vivid it may seem in other respects. The Standard Model of Particle Physics is doomed because it does not even permit an *approach* at stating a comprehensive theory of matter including also gravity. If all possible approaches to explaining gravity in terms of that model have failed, then the model itself has failed, and physics must try to find something better.

CHAPTER 12:

ON THE ORIGIN OF THE UNIVERSE

Possible visualizations of ether; original motion presupposed. I. Hypothesis of deceler-ating ether methodologically sound; sounder than assuming attracting force inherent in matter; existence of moving ether confirmed by experiments; possible experiments for testing its deceleration. II. Speculations on cosmology; speed of light in outer space unknown; calculations of time since beginning of the universe impossible; hypothesis of decelerating ether increases consciousness of human ignorance; assumingoriginal motion of ether permits theistic interpretation; reconciling critical rationalism with religious beliefs important for unity of world of thought.

The explanation of gravity put forward in the last chapter was highly con-jectural. It started from the hypothesis that ether is is in permanent mo-tion and filling the universe, and that matter itself *is* ether, though in the shape of superimposed effects. In that notion we might visualize the uni-verse of matter as a gigantic vortex resembling on an even larger scale the galaxies which we can observe more directly. Or we can try to visualize it as an avalanche rolling along in space filled with immaterial ether. In that material avalanche galaxies might be like the crystals in a snow avalanche, themselves rolling along in tiny sub-systems of clusters or vortices.

An essential feature of that approach is that it presupposes an *initial* velocity of ether, coming from outside and decelerating in the direction of the centres of every single material particle or body. This chapter will mainly be about some cosmological inferences following from that sur-mise. Before coming to that however, I must defend the approach against criticism which I expect.

I

Some may distrust the hypothesis of decelerating ether because it presupposes so much while at the same time leaving unexplained even more, most of all where ether came from and what caused its initial velocity. The objection is true but not valid. I discussed it in *Popper versus Einstein* but must repeat some aspects here because my theory would be incomplete without them[388].

(1) One aspect is at the level of methodology. We saw more than once in this essay that new theories will always raise also new questions. That follows from Popper's observation that 'scientific explanation, whenever it is a discovery, will be the explanation of the known by the unknown' [389].

In empirical science discoveries consist in introducing new hypotheses for explaining phenomena of nature, and in then finding a way of testing whether they are false or could be true. Explaining the new hypothesis itself would already be the next step in the evolution of science because it would again have to consist in explaining the known by the unknown. It is demanding too much from a new theory therefore.

The fact that the hypothesis of decelerating ether cannot explain the origin of its motion but must start from assumptions that remain unexplained is not a sign of its weakness therefore. It is stronger for instance than Newton's theory describing gravity as a force of attraction inherent in matter because that is also in need of explanation as we saw[390]. In fact, by abstract comparison the hypothesis of decelerating ether is by far the stronger of the two because assuming an attracting force of matter would explain no more than gravity itself. It would not even explain why no shield or filter can interrupt it. The hypothesis of decelerating ether, by contrast, will explain that and many other phenomena as well without

---

388 Chapter 9 (p. 177–186).
389 I quoted the text more fully in section *II, 1* of the *Introduction,* above.
390 See *Chapter 1, I,* where I quoted from Newton's famous letter to Bentley.

disagreeing with other results of Newtonian physics[391]. If we assume ether to exist, then assuming it to be at rest would be far more abstruse than assuming it to be in motion itself. Even if the origin of that motion remains unknown the hypothesis of decelerating ether will nevertheless enlarge our empirical knowledge, though only in the shape of a new hypothesis which can be true or false and which we must try to probe by experiments.

Assuming ether to be in motion *initially* makes no difference from the point of view of methodology. The hypothesis of decelerating ether does not claim to be an explanation of itself. It is a hypothesis deserving to be tested by experiment. It may seem daring to physicists believing Newtonian physics to apply only to systems at rest, but their belief arose from a confusion of mathematics and physics as we saw. The hypothesis of moving ether is a rational approach to the problem of explaining gravity.

(2) The next question is whether it will be possible to test the truth of that hypothesis in experiments. I think there can be no doubt of that even at present, but we must distinguish between the existence of ether and its deceleration for seeing that.

The experiments mentioned in *Chapter 6, III* already confirm that moving ether *exists*. One of them was that on time dilation. Transporting caesium clocks around the earth in fast aeroplanes had influenced the rhythm of the atoms controlling those clocks. After the downfall of the theory of relativity the only plausible explanation of that effect that I can think of is to consider it a long-distance effect of ether wind which thus confirmed that ether must exist. Fizeau's and Sagnac's experiments on the influence of flowing water or of a rotating source of light on the velocity of light rays also support that view. Both show that the motion of *matter* will influence the velocity of *light*. Since that would be inexplicable if light

---

391   Only the hypothesis of decelerating ether explains why the force of gravity cannot be interrupted. It also explains the transportation of light (*Chapter 11*; see also *Popper versus Einstein*, p. 158), the force of gravity itself (*Chapter 11, V*) and entropy (*Chapter 11, V,3*, conjecture *b*). Beyond that, as I showed in *Popper versus Einstein*, it also explains Eddington's experiment (p. 132), and geomagnetism (p. 134), and the spiral shape of galaxies (p. 129) and the earth's rotation (p. 154).

had no carrier, we must interpret those experiments too as confirming the existence of that carrier, therefore of ether. And the same goes for the Hubble effect which also indicates the existence of a carrier of some kind.

(3) Devising experiments for testing the *deceleration* of ether is more difficult but not impossible. Some of Michelson's measurements of the velocity of light seem to indicate already that it decelerates when approaching matter[392]. The Hafele/Keating-experiment mentioned in *Chapter 6, III* might also provide an approach if carried out at greater altitudes. And in *Popper versus Einstein* I proposed two new approaches for testing that surmise. One of them would rely on the principle of Rœmer's first measurement of the speed of light but would use measurements at different positions of the earth on its orbit. The other would consist in getting more reliable measurements of the speed of light at different altitudes[393].

If they confirm that light decelerates as it approaches matter, then it will be reasonable to assume the cause of that deceleration to lie in its carrier, therefore in ether.

(4) At the level of ether we could then find plausible explanations not only for deceleration itself but also for the vortex shape of galaxies.

We saw in *Chapter 11, V* that direct contact between elastic ether particles will explain the deceleration of ether because it is natural to assume the pressure of ether particles on adjacent particles to be stronger near the centre of matter than at greater distance from it. Given that, it is also natural to assume that pressure to slow down the elastic reactions of ether particles to impulses received from other particles. And in *Conjecture b* we assumed that torsional impulses created in a collision of ether streams will explain the phenomena of *inertia* and of of *entropy* because elasticity implies a delay of reaction to an impulse. Taken together, those two assumptions will then explain also the the vortex shapes of hurricanes here on earth and that of galaxies in the universe.

---

392 Jaffe *Michelson and the Speed of Light* (1971), p. 164ff.
393 *Popper versus Einstein*, p.189–195.

We know that inertia is a proportional to mass multiplied by velocity. If a ball moves on a stright line and at the same time rotates on an axis at a right angle to that line, then part of its mass is moving in the same direction as the ball and the other part is moving against it. The part moving in the same direction then is faster than the ball itself and the other part is slower than the ball itself. Inertia therefore is higher on the fast side of the ball than on the slow side. Assuming the air drag to be equal on both sides, its decelerating effect on the moving ball will therefore be *proportionally* stronger on the slow side than on the fast side which will cause the ball to fly in a curve with its slow side being the inner side. Billard players, tennis players and golfers will know that effect. If we assume the same principle to apply also to the vortices of which we believe ether itself to be composed, then that would explain also the the vortex shapes of hurricanes and of galaxies.

The resemblance of those vortex shapes seems to attract little attention in theoretical physics of our days. That still amazes me because René Descartes (1596–1650), living at a time when telescopes had only just been invented and when other galaxies could not yet be observed, had already considered the vortex shape to be fundamental to the whole universe. I mentioned above that at the time of writing *Popper versus Einstein* I had neither even heard of his vortex theory nor read Huygens' ether theory. That makes the coincidence of our thoughts the more striking.

The typical vortex shape can be observed everywhere in nature, here on earth as well as in space, even in water flowing through a sink. What could be more natural than assuming that nature will repeat itself? That the earth herself is a vortex and that the vortex shape of galaxies may ultimately have been caused by the principle of resonance letting their components of ether arrange themselves statistically in the direction of their torsional impulses? Or, looking at it from the other side, what could be more natural than conjecturing that rotation existing in every tiny vortex of ether will eventually result a similar rotation of the immense masses of the Earth and even in the rotation of celestial galaxies, perhaps even of the whole universe? The only plausible explanation for the vortex shape itself is that there is a deceleration of speed towards its centre[394]. The principles of inertia and resonance, both implied in the elasticity of ether, would thus explain the forces acting in smallest particles of matter as well as those in celestial galaxies, perhaps even in the whole universe.

The congruence of the shapes of galaxies in outer space and those of hurricanes and other kinds of cyclones here on earth is probably the strongest argument for assuming ether to be decelerating in the direction of the centre of matter[395]. It is obvious that the spiral shape must be caused by rotation in both cases. For the origin of that rotation I can see but two possibilities. Motion could either originate from the centre of the vortex or from its periphery. Assuming it to originate from the centre would imply a centrifugal force. At least for galaxies and cyclones it would raise the question of what prevents the particles of ether or matter from propagating away from the centre on straight lines instead of in a spiral. It could not be reconciled with Newton's First Law but would raise new problems instead.

The only explanation agreeing with Newtonian physics is that the spiral shape of galaxies and cyclones must originate in a force coming from the *periphery* of the vortex, therefore from outside, just as the force of gravity causing a vortex in a sink is coming from outside[396]. The rotation of galaxies might then have originated in a collision of ether streams which created torsional moments at every point of collision of its particles - an effect well known from almost-head-on collisions of motor cars. That surmise also conforms to the explanation of all types of cyclones occurring here on earth. Meteorology explains them by external forces because that seems to be the only plausible explanation. It is reasonable therefore to assume also the cause of the rotation which we observe in galaxies and cyclones alike to be coming from outside.

Starting from that assumption we can even explain the deceleration of ether itself as resulting from increasing pressure. We must distinguish

---

394 In *section II,* below, I will discuss that aspect in more detail. See also *Popper versus Einstein,* p. 128.

395 *Chapter 11, VI;* see also *Popper versus Einstein,* pp. 119–140.

396 *Popper versus Einstein,* p. 128; see also *Chapter 11, V.*

between the rotational speed of a vortex expressed by the number of *revolutions* per time unit and the circular speed of the particles within it expressed by the *distance* they cover per time unit. The number of revolutions per time unit may increase as a material particle is travelling from the periphery of the vortex to its centre. But the circular speed of that particle, expressed in distance covered per time unit, may nevertheless decelerate as the radius is getting smaller. Assuming deceleration is natural in that situation because particles closer to the centre of the vortex will be under higher pressure than those lagging behind. The elastic response of every single particle will impede that of the next one following it. In cyclones here on earth it may be possible to make visible that effect, and to measure it.

As a result of all that, I think we can say that the hypothesis of ether being in *motion coming from outside* and *decelerating* in the direction of the centre of matter is consistent with all empirical and methodological postulates of a critically rational approach to science in the sense of Popper's theory.

II

The hypothesis of decelerating ether can also provide a rational approach to discussing problems of cosmology. It leaves the origin of ether and the cause of its motion unexplained, but we saw above that explaining them would already be the next step in the evolution of science because it would again have to consist in explaining the known by the unknown. Man never may be able to take that step, and he hardly will be so before having probed by experiments what empirical truth there is in the hypothesis of decelerating ether[397]. But that need not prevent us from speculating about cosmology.

---

397   One of the reasons why man may never be able to take that step is that the universe seems to be a singular process; see *Popper versus Einstein*, p. 186.

(1) There are different ways of speculating on cosmology but bearing in mind how little we know is important for all of them. The origins of ether and its motion are unknown but that is not all. If we assume ether to be the carrier of light and to be decelerating as it approaches matter, then we know not even the velocity of light in outer space. That means that distances to other celestial bodies may be different from what we would expect them to be if the velocity of light were constant[398]. In fact, it means that we cannot precisely measure any distances at all in the universe as long as we use light for measuring. Instead we must assume all calculations based on terrestrial measurements of the speed of light to be mere *approximations* at best. Even that is open to doubt because there may be higher or lower concentrations of matter in other parts of the universe, and the deceleration of the speed of light may accordingly be greater or smaller. It also is possible that light becomes visible to human eyes only as it decelerates when approaching matter.

Another consequence of the hypothesis of decelerating ether is that all calculations about the time that passed since the beginning of the universe break down. There is no reason to mourn that because the only theory pretending to know something about the beginning of the universe is Big Bang theory which never was a scientific theory anyway. It could not even provide plausible explanations for all the rotation going on here on earth and in our solar system and in the whole universe. Its development compares quite well to that of quantum theory and the theory of relativity. No one ever discussed it to the end, yet almost everyone takes its truth for granted now. It may be fashionable in our time but that only tells against our time because the explanatory potential of Big Bang theory ever was poor. Basing mathematical calculations on that theory simply feigned knowledge that never was.

---

398 In *Popper versus Einstein*, I tried to calculate the speed of light on the principles which Rœmer had used, and I reached a result that is 608 km/s higher than the highest terrestrial measurement (p. 160, see also p. 189f.). My calculation presupposed however that our knowledge of the diameter of the earth's orbit is sufficiently precise.

(2) The hypothesis of decelerating ether thus not only adds to our knowledge but once again also to our consciousness of our own ignorance, which is the precondition of curiosity and therefore of all progress of empirical science. It has another advantage however, which is not at the level of science yet may be important for its progress.

If we assume ether to be in permanent motion and that motion to be coming from outside the universe of matter, then that will permit also a *theistic* interpretation not only of matter and of gravity but of all physical phenomena and of the whole universe. I am not religious myself and cannot speak for religious people therefore. Besides, there are more varieties of religion even than of physical or legal theories. Nevertheless I think most religious people could assume without difficulty that the moving ether proposed in this essay is to them the *Breath of God* which once created the world and now keeps it going. They could believe that by pressing us to the centre of gravity, He is permanently saving us from being catapulted into space by the centrifugal force of the earth's rotation according to Newton's first law of motion.

The mere possibility of interpreting the theory in that theistic sense is important in my opinion because it opens a way of reconciling with each other religious beliefs and the most rational and critical approach to scientific thinking. It may even provide a rational approach to explaining transcendence as I have shown elsewhere[399].

At present there seems to be a deep gulf between the camps of religious and non-religious people. At least my personal impression is that religious people will be averse to speaking about fundamental questions of physics more often than others, and that they will often refuse to be drawn into discussions on such topics. Other than some theoretical physicists seem to think however, I do not believe the cause of that gulf to lie in people with religious beliefs being unable to follow the ways of physics. There are too many examples in history telling against that, and we saw too many errors in present physical theory for assuming that explanation

---

399 *Popper versus Einstein*, p. 186.

to be plausible. The far more important reason is probably in the desire of religious people to remain undisturbed in their beliefs. It might originate in a conflict between their desire to be truthful and their apprehension that results reached by science might conflict with their religious beliefs. As long as they believe that the results reached by theoretical physics cannot be reconciled with believing in God, some will be unwilling even to listen to them.

Bridging that gulf would be important for the progress of science. If the approach proposed here is compatible with a theistic understanding of cosmology in which religious people can interpret the universe as a creation of God, and the motion of ether as the Breath of God which keeps the world going, then no reason should prevent them anymore from taking that rational approach to the problems of science. It is a rational approach in the sense of critical rationalism because it is critical even of rationality itself. And the more people wonder about the enigmas of nature and turn their imagination to solving them, the greater will be the hope of mankind of making new discoveries and finding better solutions one day. Bridging that gulf is important also for the unity of the world of human thought which at present is in danger of staying divided for a long time.

CONCLUSION:

A CAUTIOUSLY OPTIMISTIC OUTLOOK

'Among the real dangers to the progress of science
is not the likelihood of its being completed, but
such things as lack of imagination (sometimes as a
consequence of lack of real interest); or a misplaced
faith in formalization and precision (...); or
authoritarianism in one or another of its many forms.'

KARL POPPER (1902–1994)

Theoretical physics obstructs progress of science systematically. I. Faraday and Popper lost the same battle; theoretical physics unreachable by arguments. II. Popper's theory of science correct but application sometimes inconsistent. III. Relevance of ethics for science; mistakes of institutio.

Readers having persevered up to this point, if such there should be, will have noticed that the explanation of gravity proposed in *Chapter 12* differs in principle from anything proposed in the 20th Century. They may also have realised that it would have been next to impossible to make anyone accept that explanation without first getting out of the way quantum theory, the theory of relativity and the standard model of particle physics. I would have preferred coming to the point more directly in the whole essay and in each chapter, but it always seemed necessary first to prepare the ground by explaining the historical background of a situation and to discuss only afterwards the problems arising from it. The present situation made all other approaches look hopeless from the outset.

The main concern of this essay however was not about theoretical physics but about the theory of science, though in its application to theoretical physics. I would like to return to that now. What makes the misunderstanding we saw so serious is that it *systematically* obstructs the progress of science. The fundamentally mistaken approach taken by theoretical physics in the 20th Century still dominates today's theoretical physics. It not only is a secular error but is apt also to entail secular consequences. Theoretical physicists must decide whether to fall back mentally into the dogmatism of the Middle Ages or to make progress again by permitting scientific discoveries explaining the known by the unknown.

The great problems of epistemology have long been solved in principle. In his *Logik der Forschung* (1934) and in *Conjectures and Refutations* (1963) Karl Popper showed a solution that met with no serious opposition in the more than eighty years since he first published it. I disagree with him on some questions of its application, but as a theory of scientific discovery his solution appears to me fully convincing even after wrestling with problems of the theory of science in almost fifty years. Werner Heisenberg, incidentally, must also have seen things like that. I must say that again to all physicists. Those admiring him as a theorist should not carelessly put aside thoughts he took over from Karl Popper almost *verbatim* even if reproducing them as his own[400]. Relevant counter-arguments to Popper's approach to the theory of scientific knowledge are unknown to this day.

Karl Popper lost a great battle nevertheless. It was the same battle that Michael Faraday also lost but did not want to give up to the last, the battle for an imaginative science, a science not limited to what we see with our eyes but searching for the reality of things that are invisible to human senses.

Faraday practiced that imaginative science like none other, and he achieved unprecedented success with it. His discoveries changed the whole world. Had he not been so firmly convinced that behind the visible phenomena of the world there must be also another world of invisible phenomena, then electrical induction might never have been discovered.

---

400   See the footnote at the end of *Chapter 8, II.*

Without induction there would be no generators and no electric motors, no power plants, no electric light, no petrol engines with electric ignition, no telephones, no radios or television sets and no computers. The evolution of technology would have ended with the steam engine. Without Michael Faraday there would have been no technological revolution of the 20th Century.

If human lack of imagination were not as boundless as it sometimes is, then Faraday's achievements should actually have convinced everyone that science must be *speculative* if it is to make discoveries. Even more should this have happened after Popper had provided the theoretical understanding by showing that scientific explanation, whenever it is a discovery, must be the explanation of the known by the unknown, and that science does not proceed systematically therefore, but in revolutionary leaps, inventing theories first and probing them only afterwards by the standard of reality.

Neither Faraday nor Popper however could reach with their teaching the proponents of that discipline to which their thoughts were most strongly devoted. They could not reach theoretical physics. Many theoretical physicists still persist in believing theoretical physics to be an 'exact science', seeing its objective only in describing and making calculable as precisely as possible what ostensibly we know for certain. They fail to see that physics would never have reached its present level if our ancestors had also taken that attitude. And they fail to realise that by taking it, theoretical physics would cut itself off from making discoveries, and thus from all future development.

Karl Popper, I fear, had himself to blame for that defeat to some degree. His theory of science was correct but his application of it was not always consistent. He had found the right approach but he did not lay the axe to the root. The reality behind appearances for which he was searching was no physical reality to him. As to Planck and to Einstein it still was the reality of mathematics. He never faced the question of whether theories trying to find *ultimate* explanations in mathematics, such as quantum theory or the theory of relativity, were compatible with his own criterion of falsifiability. I will show that in more detail in *Appendix 1* of this essay and will propose an explanation for Popper's error there. That error does not affect his theory of science however, but only some conclusions he drew from it in the individual sciences.

Applying Popper's theory to the individual sciences is more difficult than I expected. That makes it so interesting. It means fighting against prejudices rooted deeply and firmly since antiquity in the traditions of thought of the Western world. Those prejudices are not only in others but even in ourselves, in every one of us, including myself. Karl Popper like none other fought them to the end of his life. He did so in his *Open Society* and in all his works. I fear, however, that they will prove to be ineradicable. Like the Hydra of Lerna they will continue to grow new heads, today and in future and in all branches of science. For that, too, Popper is the best example. Although by far the greatest philosopher since Kant, he nevertheless sometimes contradicted himself in the application of his own thoughts. At the bottom of his mistakes there always was an insufficient implementation of the Kantian principle of autonomy. I hope that others will not only improve on my mistakes but will recognize also the sincerity of my endeavours.

APPENDIX 1:

ON KARL POPPER'S THEORY OF SCIENCE

> ,A rationalist is simply someone to whom learning is more important than being right; someone willing to learn from others, not by taking over their opinions simply but because he likes having his own ideas criticized by them and criticizing theirs.' [401]
>
> KARL POPPER (1902–1994)

Karl Popper shaped my views on the problems of the theory of science more than anyone else did. Nevertheless they are not the same as his in all respects. First differences showed in my '*Recht und Rationalität*' (1984), others while I was writing *Popper versus Einstein* (1998), some only when working on *Albert Einstein oder Der Irrtum eines Jahrhunderts* (2009), and the most recent even when working on this book and its appendix. Consistently applying Kant's principle of autonomy is at the bottom of all of them.

Popper would have wanted me to state those differences as clearly as possible in order to permit discussing them. I know that because in his last year he even took steps for initiating a discussion on some of them. Only his final illness prevented it[402]. The objective of this appendix is to encourage that discussion after his death and mine.

---

401 My translation from Popper, *Alles Leben ist Problemlösen*, p. 160.

402 At Popper's instigation, the University of Madrid invited me to read a paper at a summer seminar, which it later cancelled it due to his last illness. In 1995, I read a new paper on the same subject at a conference at Prague (see following note).

I

Some of the differences between Popper's views and mine show clearly in their results. That particularly applies to quantum theory and the theory of relativity. He agreed with both in principle whereas I believe both to be fundamentally mistaken in their approaches as well as in their results. We saw that in *Chapters 8* and *9*. Other disagreements are about details. One of them concerns the relationship of facts and norms, not relevant for this book. Popper assumed a 'critical dualism' of facts and norms and believed that to be fundamental whereas I believe it to be inconsistent with other views of his, but also to be irrelevant[403]. Our disagreements on the theory of logic and mathematics, mentioned in *Chapter 4*, seem also to concern only details but causes lie deeper there. They make themselves felt also in other fields, for instance in that of the theory of probability.

In all cases I see no disagreements in our notions of the theory of scientific discovery itself but only on applying it correctly. I think Popper's application of Kant's principle of autonomy was not radical enough and therefore not always consistent. I emphasize once more, however, that this in no way diminishes my admiration of Popper and his achievements. If there is any merit in this book I owe it to him, and I deeply regret that my interest in theoretical physics arose too late for discussing with him its problems. And as to understanding Kant, it is due to Popper alone that I could ever believe having understood him, let alone develop a view of my own on his theory.

Many philosophers never noticed that Kant's critical theory had been a revolution in the theory of knowledge. Overlooking that is not rare even among his followers. Some of them completely neglected his groundbreaking discovery of the principle of autonomy but successfully spread his errors, doing much harm thus to his reputation and to the intellectual

---

403 I explained that in *Recht und Rationalität* (1984), p. 35–43, and in *The Problem of Objectivity in Law and Ethics* (1999).

climate of Kantianism[404]. Popper however not only understood Kant but also was always generous in interpreting him. That alone deserves greatest praise. He never tried to place himself above Kant and prove the greater philosopher but always was interested only in the issue in question and in making the most of Kant, if possible even more than Kant himself had done. He took up and brought forward Kant's revolutionary doctrine of autonomy; and he analyzed and tried to mend the mistakes which Kant had been unable to avoid. I repeat that without Karl Popper I would never have been able to shape a view of my own on Kant's theory.

But although Popper explained Kant's principle of autonomy more convincingly than anyone before him, more convincingly even than Kant himself, I nevertheless think he did not apply it consistently. In questions of epistemology I see no difference between his view and mine, but in his application of his own theory I see a breach going through most of his works relating to physics. At any rate, my own position is even more radical than his was.

That implies that I am criticizing Popper in a constructive spirit. I believe that *in its application* the theory of science must become more 'Popperian' than even Popper himself had stated it. That is what I am going to suggest here.

(1) On the principle of autonomy: In *Chapters 4* and *5* I explained my view of the bearing which Kant's principle of autonomy has on the understanding of reality and morality, and that it applies also to the language we use, and therefore to *logic and mathematics* which are part of that language. My impression is that Popper either did not see that aspect or did not realise its importance. That must have been why he

---

404 I particularly have in mind such books as, for instance, G.E. Moore's *Principia Ethica* (1903), or Hans Kelsen's *Reine Rechtslehre* (1934 – 'Pure Theory of Law'). Both claim to be standing in the Kantian tradition, but both are concerned almost exclusively with investigating the *true* meaning of concepts. Popper must have been thinking of that kind of philosophy when he spoke of the 'empty verbiage and barren scholasticism' of the essentialist approach to philosophical problems (*Open Society*, vol. 2, p. 9), and when he claimed that for discussing normative problems, we should use the 'language of political demands or of political proposals' (*Open Society*, vol. 1, p. 109).

did not apply consistently the principle of methodological nominalism introduced by Alfred Tarski.

In Popper's theory logic and mathematics belong to his 'World 3' containing the products of the human mind. He always emphasized their being part of reality[405], and I share that view without reservations. However, I regard neither logic nor mathematics as being *closed systems* but as being part of language which also is no system but open in principle. Man created language. It lives with him, and he shapes it. Unlike empirical truth it is not independent of man therefore. Nevertheless it may raise objective problems and may even contain objective truths. They are only *intrinsic* problems or truths however, referring always only to language itself, and never to empirical reality beyond it. The connection between language and empirical reality can only be established by assigning to the terms we use in our language a meaning going beyond those terms themselves, letting them thus stand as *symbols* for some object. Choosing that connection by deciding which term is to stand for what object is an act of the autonomous human mind, and the standard by which we must gauge it is not the standard of truth therefore but that of *expediency*.

I believe Popper must have held the same view, but am not sure of that[406]. I see no fundamental difference between his theory of scientific knowledge and my own view, only a difference of opinion with respect to consistency. However, differences of that sort, seemingly insignificant at first glance, may sometimes lead to disproportionate consequences. In our assessment of quantum theory and the theory of relativity the dif-

---

405 See Popper, *Objective Knowledge* (1973, p. 123 f., 172 f.; Popper/Eccles, *The Self and its Brain* (1977), section 11. The beginnings of his doctrine of 'Three Worlds' appear already in *The Open Society and its Enemies*, vol. I (1945), ch. 5, III (p.65), also in Appendix I of *The Open Universe An Argument for Indeterminism* (written before 1956, published 2001).

406 Popper/Eccles, *The Self and its Brain* (1977). In section 11, Popper uses an example taken from mathematics and assumes statements like '2 x 2 = 4' to contain a truth that is 'invariant' to conventions and translations'. – My view, by contrast, is that theorems like 2 x 2 = 4 have nothing to do with truth, and that their invariance to conventions and translations is due to their being non-empirical statements that do not refer to a reality independent of man.

ferences between Popper's view and mine obviously are unbridgeable in their results. They are not due to a theoretical disagreement in matters of principle however, but mainly to a difference in the awareness of problems. Some of the questions raised in this book should actually have been problems for Karl Popper too, but his understanding of mathematics and logic, and probably also the gratefulness he felt for Einstein, prevented him from seeing that, and let him pass over them.

(2) Popper's ambiguity on the role of mathematics: In my opinion Karl Popper overtaxed the role of mathematics in most of his works on questions of theoretical physics. He sought in the rules of mathematics *ultimate* explanations for physical phenomena without realizing in that context that mathematics can provide no explanations but only descriptions, and that in empirical science there are no ultimate explanations anyway.

Popper intended his criterion of falsifiability to serve as the demarcation between empirical science and metaphysics but his attitude towards mathematics was contradictory. That is understandable when taking into account the times in which he grew up. When he was born in 1902, Max Planck had just devised his quantum theory (1900) and the inauguration of the theory of relativity (1905) was imminent. His entire youth thus stood under the influence of the new type of mathematical physics, then hailed as a wonderful advancement of humanity, and doubted not even by professionals. What I criticize as a fundamental confusion of mathematics and reality in the essay must have appeared to him the epitome of modern science. How could he have questioned it with his then still unfinished beliefs? What standards of criticism should he have employed? Young scientists lacking experience depend on principles far more than old scientists. Popper too could reach only with difficulties his own critical attitude. We must not be surprised therefore if he did not succeeded in one step and not in all fields at once, especially if we ourselves had the good fortune of becoming acquainted with his thoughts early in life. Applying them to quantum theory and the theory of relativity is much easier under such conditions than it must have been for him.

In this appendix, however, I am discussing questions of the theory of science. And in that field no human sympathy and not even my boundless admiration for Karl Popper can alter the fact that his attitude towards ether theory and towards the empirical content of quantum theory and the theory of relativity was inconsistent and ultimately mistaken. Going by his own criteria he should have rejected from the outset both quantum theory and the theory of relativity, and instead been an ardent supporter of the ether hypothesis. The ether hypothesis has every feature that a good physical theory must have in his theory of science. It describes a reality *behind* visible phenomena for which, as he said, any true scientist must search[407]. It introduces as a daring hypothesis important new empirical information for explaining the known by the unknown. And it can so be stated that the inferences derived from it are accessible to experimental tests and falsification. We saw that in *Chapter 12, I, 2.* Nevertheless, in the shape of Fresnel's hypothesis of moving ether, it withstands all experimental tests still today, as I have shown in the essay and elsewhere[408].

In other cases Popper made a principle of interpreting benevolently a theory and making it as strong as possible before criticizing it. Had he applied that principle also to ether theory, then he might have reached the conclusion that ether is not something distinct from matter, but that matter is a superimposition of ether as I have suggested in *Popper versus Einstein* and in the essay[409]. He might even have found his own vision confirmed

> 'that there are both particles and waves and that material particles will be controlled by immaterial waves'[410].

---

407   He thought that very important. See Karl Popper, *Kepler Seine Metaphysik des Sonnensystems und seine empirische Kritik*' in Popper, *Alles Leben ist Problemlösen*, p. 145 f., 146; also *The Myth of the Framework*, p. 116; *The World of Parmenides*, p. 106; Popper/Eccles, *The Self and ist Brain*, section 3.

408   *Popper versus Einstein* (1998), Chapter 6 ff.

409   In *Albert Einstein oder Der Irrtum eines Jahrhunderts*, my German text here says (p. 278) that ether is a superimposition on matter (instead of *vice versa*). I noticed that mistake only when making the translation.

410   My translation from Karl Popper, *Alles Leben ist Problemlösen* (p. 148); see also

For that, I think, is exactly the view to which the ether hypothesis will lead if carried through to its end[411].

Popper followed a different approach however, and in doing so misjudged the relevance of mathematics. Quantum theory and the theory of relativity stand in direct opposition to his theory. Both place supposedly safe logical and mathematical operations in the place of daring empirical conjectures. And both lead to ultimate explanations not open to further discussion any more. For that very reason those logical and mathematical operations can only be either boring or mistaken. If they observe the rules of logic and mathematics, then they must consist in tautologies and therefore be boring because any computer could do better. And if they do not consist in tautologies, then they must violate the rules of logic and therefore be mistaken.

From his own point of view Popper discarded both theories from the outset. In both cases his blurred vision of logic and his divergent understanding of mathematics, which he regarded as a *valid system*, caused him to be impressed too strongly with the mathematical side of those theories, and to neglect the question of their empirical content.

(*3*) *Quantum theory*: That shows already in the way Popper treated quantum theory. His book *Quantum Theory and the Schism in Physics* was written in 1951–1956 as Volume 3 of his *Postscript to The Logic of Scientific Discovery* (1934) but published only in 1982. He expressed very clearly there his deep uneasiness about the development of quantum theory, especially since the so-called 'Copenhagen interpretation of quantum mechanics' (1927). That deserves greatest admiration in my opinion. Popper foresaw the crisis of theoretical physics at a time when most still believed it to be at the height of its achievements.

Popper versus Einstein p. 119 ff.

411 The word ‚immaterial' has nothing to do with ‚spiritual' in this context; I think the sense Popper had in mind must have been closer to that of Hugyens' theory explaining the transportation of energy without transportation of matter only through the elasticity of the transporting medium (see *Chapter 6, I*).

Popper's criticism falls short of the problem however. The question of whether or not quantum theory is apt to solve *any physical problem at all* simply does not come up in his treatise. He apparently never asked himself that question although everything depends on it. That must also have been why no discussion can be found anywhere in his treatise of the empirical relevance of quantum theory, that is of the question of whether it can be tested experimentally and has withstood such tests.

Popper addressed epistemological questions of physics in detail again in his contribution to the treatise *The Self and Its Brain*, published in 1977 together with John Eccles. Even there, however, he did not mention the problem of the empirical aspects of quantum theory. He would hardly have done that on purpose, but he must have lacked awareness of the problem. He simply never seems to have questioned that quantum theory is an empirical theory. Consequently, he also never seems to have asked himself whether there is any empirical evidence for the demanding statements of that theory, especially for the allegation that there is a fundamental and indivisible constant of nature of the size of h.

In *Chapter 8* I explained why I consider quantum theory to be mistaken in its approach already, and why it was by no means an empirical theory, at any rate not as Max Planck had stated it. Popper however seems not even to have asked himself whether it is open to experimental tests at all. Since this attitude cannot be reconciled with his own theory of science as explained in his *Logic of Scintific Discovery*, I can explain it only by assuming that he had not seen the problem. That in turn must have been because he was still relying on the axiomatic understanding of science as represented by Planck and Einstein in such matters although it is incompatible with his own theory. That is why I think that although beyond question in their results, the differences between Karl Popper's theory and mine nevertheless are not fundamental. They do not concern his theory of science but only its application to theoretical physics. Our disagreement might have been fundamental if Popper had explicitly discussed the question of the falsifiability of quantum theory, and then had reached a different result. Since he did not discuss it at all however, the disagreement is not on the theory of science but only on the correct way of applying it.

(4) Theory of Probability: Popper's theory of probability refers to quantum theory and raises another question concerning it. Popper thought he could solve the methodological problems of quantum theory by postulating the *physical reality* of fields of propensities which should then serve as the cause of the quantum theoretical indeterminacy units[412]. Explaining that view he wrote in *Quantum Theory and the Schism in Physics:*

> 'I believe that the waves (even those of the second quantization) are mathematical representations of *propensities*, or of dispositional properties, of physical situations (such as the experimental set-up), interpretable as propensities of the *particles* to take up certain states[413].

> One must remember that, according to the interpretation proposed here, propensities, and fields of propensities, are just as real as forces, and fields of forces. They are, like forces, dispositional properties; and like forces, or fields of forces, they are properties not so much of the particles as of the total physical situation; they are, like forces, *relational* properties'.[414] (Popper's italics).

The text shows that Popper considered the 'propensity' of a situation, that is to say its tendency to bring about certain physical conditions, a physical property of the situation itself. He called it a 'propensity field', which in turn he regarded as a physical property of an experimental set-up, or an overall arrangement[415]. At any rate, that is how I understand him. He wanted to raise probability itself to the rank of a physical property that was to serve as a physical explanation of indeterminism.

In my opinion Popper did not solve the problem of quantum theory by that theory but carried to the extreme the mistake standing at its begin-

---

412   Karl Popper, *Unended Quest* , ch. 34 (p. 151 f.), S. 221 ff; *Realism and the Aim of Science*, S. 347 ff; *Quantum Theory and the Schism in Physics*, p. 42 ff., 79 ff., 144 ff., 151 ff.; *The Open Universe An Argument for Indeterminism*, p. 93 f..

413   *Quantum Theory and the Schism in Physics*, p. 126.

414   *Quantum Theory and the Schism in Physics*, p. 127.

415   *Quantum Theory and the Schism in Physics*, p. 127 198, 202, 204.

ning. Before explaining in detail that view I must mention that I accept one of the results of Popper's *propensity interpretation* and believe it to be very important. I think we can make meaningful statements on probabilities not only where large numbers of uniform events are involved but also about singular events. We can even make *meaningful statements on quantified probabilities of singular events*. That is the point where any mathematical interpretation of the calculus of probability will meet with difficulties because it must assume that statements on probability are meaningful only if referring to events that happen more than once.

My criticism of Popper's propensity interpretation does not set in at the point of singular events therefore. Rather on the contrary, I see things like this. If, for instance, a soccer match between teams A and B is coming up, then we can discuss before the match has begun what its outcome is likely to be. Up to its end we can also talk quite reasonably about how it will *probably* end. Bookmakers even make money on that. The same principle applies to physical situations or to the experimental setups discussed by Popper. To that extent I agree with him.

However, the fact that we can rationally discuss its likely outcome before a match has begun is not due to any physical properties of the soccer players, or of the teams or the referees, or of the court. It is not due to any physical properties at all. Once again we must solve the problem of analyzing statements of that sort by applying the principle of *methodological nominalism*. When using the word 'probability' in the context of an impending singular event we must first know what we want to say by it and others to understand by it. If we do not know what we want to say by it, then we should avoid using the word. At least in science that should be our attitude.

I believe that when speaking in categories of probability of singular events most people make use of a kind of fiction. To stay by the example: we know that the match between teams A and B will be on only once and can have only one outcome therefore. Nevertheless we can treat the situation fictitiously *as if* a large number of encounters were to be expected. We know by experience that soccer teams do not always play the same, and that not only players will influence the outcome of a match but also chance, or weather conditions, illness, injuries, or the daily form of some of the players. By fictitiously assuming a large number of encounters we can base on that knowledge the hypothesis that one of the teams will win more often than the other. And for that assessment, which consists in fictitiously assuming a large number of encounters, we can use the term 'probability' simply for making a long story short.

That, I believe, is the kind of situation most people have in mind when speaking in terms of probability of singular events[416]. Using the term 'probability' like that is reasonable too, because it expresses something we want to say and others will understand, and in which they might even be interested. The set of facts however, to which we apply the term 'probability' in such cases, is no physical property of any thing or of any situation. It cannot *explain* anything at the level of physical facts therefore. It can neither explain why in a particular case team B won the match although team A had the better players, nor why in a large number of encounters team A is not victorious every time but maybe only in 80 % of them. Probabilities are unsuitable *in principle* for serving as *explanations* of physical events. In another context Popper himself saw things like that[417].

Nevertheless using the term 'probability' in that sense does not rely on subjective interpretation alone. It is not mere 'opinion'. Being fictional the large number of events it presupposes is non-existent of course, but non-existent events too can be assessed objectively. Political economists, for instance, permanently take that approach and not only they. Other fields of research, such as archeology or epidemiology, also depend on discussing with great intensity purely hypothetical scenarios. In fact, dis-

---

416 Fictions of that kind are common in the humanities. Economics and history often make use of them for example, and the same happens in law. In consequence of Kant's principle of autonomy, applying the law means everyone must make his own legislator, and fictitiously assume a legislator endowed with reason therefore. Everyone knows, however, that in democratic states no such individual exists in reality, and that the wording of a statute is often the result of a compromise. I explained hat in more detail elsewhere (*Recht und Rationalität*, p. 106 ff; also *The Problem of Objectivity in Law and Ethics*, p. 111 ff., 118 ff.).

417 Karl Popper, *Quantum Theory and the Schism in Physics*, p. 60f.; see also my ‚Popper *versus* Einstein', p. 126 f., 211 f.

cussing fictitious events is quite common also in science. The important thing is not to confuse 'certainty' and 'objectivity', or 'subjectivity' and 'uncertainty', because those terms will normally stand for different sets of facts and distinguishing them is not difficult. We may have little reliable knowledge of the state of affairs in some matter but we can nevertheless objectively criticize and assess its uncertainty. Tasks of that sort are common in the daily work of most disciplines, be they science or non-science. In the same way we can criticize objectively also a fictional state of affairs although the word 'uncertainty' would obviously be misapplied in fictitious cases[418]. What matters is that the standard of assessment is outside man and independent of him, and that others too can apply it or criticize it. That is what constitutes its objectivity. In any case, making use of fictions is permissible in common language. In my opinion it is permissible even in science.

Fictional situations are inept for *explaining* physical events however, because they are not part of physical reality. They can 'neither kick nor be kicked'[419]. Popper's attempt at solving the problems of quantum theory by referring to propensity fields or probability fields therefore contradicts the principles of methodological nominalism. Following Einstein's example he was in fact shifting the meaning of the word 'propensity', which usually stands for something not yet exsting but likely to happen, to designating somethin existing in physical reality. Had he been right, then his explanation would have been an *ultimate* explanation permitting of no further questions. But according to his theory there can be no ultimate explanations of that kind in physics[420].

---

418 For more details, see v.Mettenheim, *The Problem of Objectivity in Law and Ethics*, sec. III, 2.

419 The expression goes back to A. Landé. Popper often quoted it (*The Open* Universe, p. 116; Quantum *Theory and the Schism in Physics*; p. 54). For his use of the term ,real' see also Popper/Eccles,*The Self and its Brain*, section 4.

420 Popper's attempt at connecting the problems of probability with radioactive decay somehow (*Quantum Theory and the Schism in Physics*, p. 95) does not make any difference to this. On the contrary, being unconnected with the problem, it appears to me particularly symptomatic of his misguided approach. For his reference to fields of propensity, too, does not tell us anything we had not already known before; it does not increase our empirical knowledge. We learn nothing new about the world in which we live, but only

(5) *Theory of Relativity*: Having discussed Popper's understanding of logic and mathematics above in section (2), and having explained in detail my own view of the theory of relativity in *Chapter 9* of the essay I can be brief here.

Popper accepted Einstein's empiricist interpretation of geometry. So he must have believed the theory of relativity to be an empirical theory. I think his mistake originated in his hazy theory of mathematics and in a lack of awareness of the problem. It is particularly understandable in this case because an idea of Einstein's had instigated Popper's own theory[421]. To some extent, I believe, the gratitude he felt for Einstein must have kept him from being more critical of Einstein's views.

The theory of relativity cannot be considered an empirical theory however, for the simple reason that it started from contradictory premises which it transposed into contradicting equations, as we saw in the *Introduction* and in *Chapter 9, IV*. Popper himself demonstrated convincingly that from contradicting premises we could derive any conclusion whatever[422]. That alone implies that we can refute the theory of relativity only by logical arguments but never *empirically*. Whatever the result of an empirical experiment may be, a skilled mathematician relying on Einstein's contradicting premises will always be able to demonstrate that this result, too, agrees with the theory of relativity. According to Popper's criterion of falsifiability, the theory of relativity therefore does not refer to reality.

The premise of the constancy of the speed of light however is either empirical or non-empirical, depending on whether we interpret it as applying *only or not only* in a vacuum. Assuming it to apply *not* only in a vacuum it is an empirical statement that was clearly refuted in Sagnac's experiment. And assuming it to apply *only* to a vacuum it is non-empirical because an absolute vacuum does not exist, at least nowhere within reach

---

some new words. That means that it sets an end to rational discussion at this point. If we explain the possibility of chance by assuming the existence of fields of propensities, then nothing remains to be explained after that. I cannot accept that, and I think from his point of view, Karl Popper should not have accepted it either.

421 Popper, *Unended Quest*, p. 38.

422 Popper, *What is Dialectic?* in *Conjectures and Refutations*, p. 317–321.

of humanity. In that shape the constancy of the speed of light is but one of those 'ideal' theories which in fact reduce to an abstract mathematical theorem an ostensively empirical statement by narrowing down to zero the conditions of its applicability[423]. We saw in the essay that Max Planck also believed in that method. Popper apparently did not know Sagnac's experiment. He hardly had a chance of knowing it because relativistic literature keeps it a secret. I am convinced that had he known the outcome of that experiment he would have rejected the theory of relativity altogether, both special and general.

(6) *Reality of time*: I discussed in *Chapter 9, V* the question of whether or not time exists, that is to say, whether it is part of physical reality, whether it has a direction and whether there can be such things as time moving backward or things moving backward in time. Popper commented in detail on such issues in a paper written in 1965 but published only posthumously under the title *Beyond the Search for Invariants*[424]. He was critical of the realistic theory of time there, but to my mind his comments neither were critical enough nor clear enough. Once more the discrepancy of our views is rooted in methodological nominalism which in turn is but an application of Kant's principle of autonomy, as we saw in *Chapter 5, II*.

In his essay Popper described space and time as being 'abstractions' and then went on to say[425]:

> 'Time, whatever it may be, becomes like space at least in so far as it forms just one dimension in a four-dimensional manifold which we may call 'objectively co-present', in the sense that no part comes objectively before or after another.'

---

423 *Chapter 7, II.*

424 That paper is now in the volume *The World of Parmenides* (1998), p. 146–222.

425 Popper said 'A realist may be doubtful about calling this manifold 'real'; for clearly, space and time are abstractions (p. 167); the quotation following in my text is from p. 170.

His confusion of mathematics and reality resembled the one we noticed already with Planck and Einstein. The very notion of a 'four-dimensional manifold' shows that like Einstein, Popper did not distinguish clearly between physical reality and the abstract coordinates of a geometrical system designed for *describing* it. His further thoughts in that paper get lost in problems of entropy which in my opinion have nothing to do with the question of the reality of the time.

(7) *Dualism of facts and norms*: Karl Popper believed in a 'critical dualism of norms and facts'. In his *Open Society* he considered that very important[426]. The different evaluations of methodological nominalism in his theory and mine affects also that issue.

My objection to Popper's critical dualism it is that it starts from the wrong question. By assuming a 'dualism' he was tryig to answer a 'what is?' question, the question of what norms or facts 'really' are, or whether or not norms are facts. In the same book, in the context of explaining methodological nominalism, he himself strongly objected to that type of question in science[427]. The aim of his question also remains in the dark because he never doubted that norms, even the norms of ethics, are 'real' in the sense of affecting the behavior of human beings and thereby causing changes in physical reality. Unless we believe that some other authority, such as God, or Nature, had given those norms they certainly would fall in his *World III* containing the products of the human mind. Only that would agree with Kant's principle of autonomy. I cannot see either that Popper ever drew from his doctrine of critical dualism any conclusions in his later works. Nothing depends on it. That is why I consider it inconsistent with his other views but also as unimportant. I have explained that in detail elsewhere and mention it here only for the sake of completeness[428].

---

426 *The Open Society and its Enemies* (1945), Chapter 5, particularly section III (vol. 1, p. 62–66).

427 Karl Popper, *The Open Society and its Enemies* (1945), vol 2, p. 9ff.

428 *Recht und Rationalität* p. 35 f.; *The Problem of Objectivity in Law and Ethics* in Jarvie/Pralong, *Popper's Open Society after 50 Years'* S. 111 ff.

II

Summarizing the discrepancies of our views, I can say that I agree with Karl Popper's theory of scientific knowledge in all points without any restraint, but that I do not always agree with the way he applied it to problems of the individual sciences. In my opinion he did not consistently apply Kant's principle of autonomy to language itself, in particular to mathematics and geometry which are part of language. None of this however prevents me from considering him the greatest philosopher since Immanuel Kant.

APPENDIX 2:

ON DEDUCING THE LORENTZ TRANSFORMATION

> 'It would not surprise indeed if time threw the hourglass in the face of such a rogue.'
>
> GEORG CHRISTOPH LICHTENBERG (1742–1799)

This appendix is for readers wanting to consider in more detail the mathematical aspects of Einstein's Special Theory of Relativity as stated in his paper *On the Electrodynamics of Moving Bodies* (1905); others may find it uninteresting. Reading it is unnecessary for someone convinced already by the arguments put forward in the essay.

In *Chapter 9, IV* of the essay I showed that Einstein's equations (1), (3) and (4) imply a logical contradiction, and that he relied on those equations in his further deduction of the Lorentz transformation. Some may argue that even if Einstein made a mistake in his time the theory of relativity nevertheless is correct because other derivations of the Lorentz transformation avoid that mistake. I contend that objections of that kind are untenable on logical grounds. The mistake may occur at different stages of the deduction and may be hidden more or less well. In any derivation of the Lorentz transformation however, there must be a logical mistake at some point because the result itself, consisting in the allegation that the speed of something may be the same relative to a system at rest as to a moving system, is contradictory. The following ought to show that. I will demonstrate there that two deductions of the Lorentz transformation, both well-known but quite different from one another, also rely on contradictory statements[429].

---

429 In *Albert Einstein oder Der Irrtum eines Jahrhunderts* I demonstrated that for three

(*1*) In his book *On the Special and the General Theory of Relativity* (1917), Einstein introduced the following *Figure 4:*

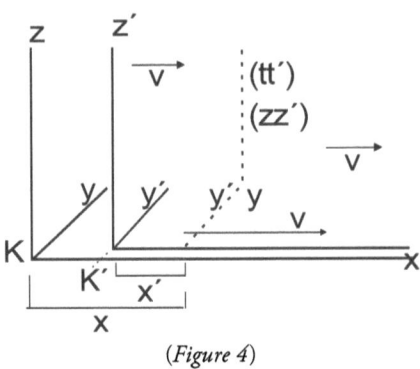

(*Figure 4*)

He explained it like this[430]:

> 'For the relative orientation of the co-ordinate systems indicated in (Fig. 4), the x-axes of both systems permanently coincide. In the present case, we can divide the problem into parts by considering first only events, which are localised on the x-axis. Any such event is represented with respect to the co-ordinate system *K* by the abscissa *x* and the time *t*, and with respect to the system *K'* by the abscissa *x'* and the time *t'*. We require to find *x'* and *t'* when x and t are given.
>
> A light signal, which is proceeding along the positive axis of *x*, is transmitted according to the equation

$$x = ct \quad \text{or} \quad x - ct = 0 \tag{7}$$

different derivations.

430 *Figure 4* quoted from the authorised translation by Robert W. Lawson of Einstein's paper in *The Special and General Theory* (1920), published by Methuen & Co Ltd.

Since the same light signal has to be transmitted relative to K' with the velocity c, the propagation relative to the system K' will be represented by the analogous formula

$$x' - ct' = 0 \tag{8}$$

Those space-time points (events) which satisfy (7) must also satisfy (8). Obviously this will be the case when the relation

$$(x' - ct') = \lambda \ (x - ct) \tag{9}$$

is fulfilled in general, where λ indicates a constant; for, according to (14), the disappearance of $(x - ct)$ involves the disappearance of $(x' - ct')$.'

Einstein then introduced the hitherto unknown constant m, and described the corresponding process along the negative x-axis by equation

$$(x' + ct') = m \ (x + ct) \tag{10}$$

'For convenience' he then introduced by equations

$$a = (l + m) \, / \, 2 \quad \text{and} \quad b = (l - m)/2 \tag{11a,b}$$

the new the constants a and b in place of the constants λ and μ and formed the equation

$$a^2 = \frac{1}{1 - v^2/c^2} \tag{12}$$

(2) Equation (12) is an important formula of the Lorentz transformation, supposedly describing the functional relations between distance, velocity and the speed of light. From the point of view of mathematics Einstein's derivation of that formula raises two fundamental questions.

(*a*) Comparing equations (7) and (8) with *Figure 4* shows that equation (8) was not correct but should actually read

$$x'- ct' = 0'. \tag{8a}$$

In *Figure 4* the zero point of the moving system K' is not the same as that of the stationary system K. Einstein had set up the equation $x'- ct' = 0$ (8) only for the *moving* system K'. By using 0 instead of $0'$ as the zero-point in equation (9) however, he transferred it to K, the stationary system for which he had not established it. In order to clarify the difference and keep it in mind I will continue to use 0 for designating the zero point of K, and will use $0'$ for designating the zero point of K'. According to figure 4 that will give us

$$0'= vt \qquad \text{or} \qquad x - x'= vt \tag{13a,b}$$

and thus for all cases satisfying $x \neq 0$

$$0 \neq 0'. \tag{14}$$

This difference originates in equations (7) and (8), which were set up for different systems of coordinates, K and K'. It disappears in Einstein's equations (9) and (10) because he introduced the hitherto unknown constants μ and λ there.

(*b*) The logical contradiction resulting from that does not catch the eye immediately because Einstein at the same time also introduced an *undisclosed multiplication by zero*. Inserting equations (7) and (8) and (8a) into (9) gives us

$$0' = \lambda \times 0 \tag{15}$$

From equations implying a multiplication by zero we can derive by seemingly correct inference any result whatever as everyone knows. In Einstein's equations (9) and (10) any value whatever could be inserted for λ and m without the equations getting inaccurate thereby; and the same therefore holds also for replacing a by some random value in equations (11a) and (12).

(2) As a result of all this we can note that the formula

$$t' = t \times \frac{1}{\sqrt{1 - (v/c)^2}} \tag{16}$$

describing in special relativity the ratio of the time in the system at rest to the time in the moving system, can be derived correctly from a system of equations treating the velocity of light as a variable, but that it cannot be derived correctly from a system of equations treating the speed of light is as a constant.

II. The Mathematical Mistake in A.P. French's M.I.T. Textbook

The following criticism refers to a US standard textbook on the Special Theory of Relativity by A.P. French[431].

(1) French explains the deduction of the Lorentz-Einstein transformation like this:

---

431   A. P. French, 'Special Relativity', M.I.T. Introductory Physics Series, (1st ed. 1968, 16th reprint, 1997), pp. 76. I published that criticism (in English) in my *Albert Einstein oder: Der Irrtum eines Jahrhunderts* (2009). In this book I have tried to improve my English but made no other changes. Only the numbers of figures and equations have been adapted to the numbering of this book.

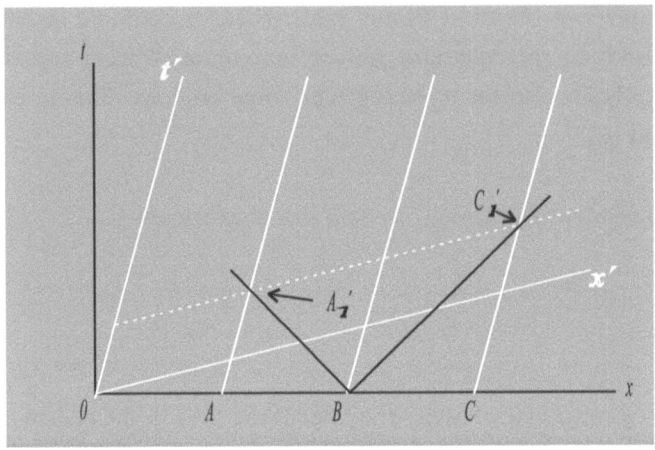

(*Figure 5*)

„Look now at (*Figure 5*). It depicts the operation of defining simultaneity at stations A and C which are moving at speed v with
respect to an inertial frame S. We have already discussed such a
diagram (...)[432]. But now we have added lines to represent the
coordinate axes of the frame S' in which A and C are at rest. How
have we done this? The axis of t' is readily described; it is the line
$x' = 0$, i.e., the world line of the origin of S'. And since the frame S'
has a speed v along the x direction with respect to S, the position
of this origin is described in S by the equation $x = vt$ if the origins
of S and S' coincide at the time $t = 0$.

---

432 At this point French refers to a previous diagram.

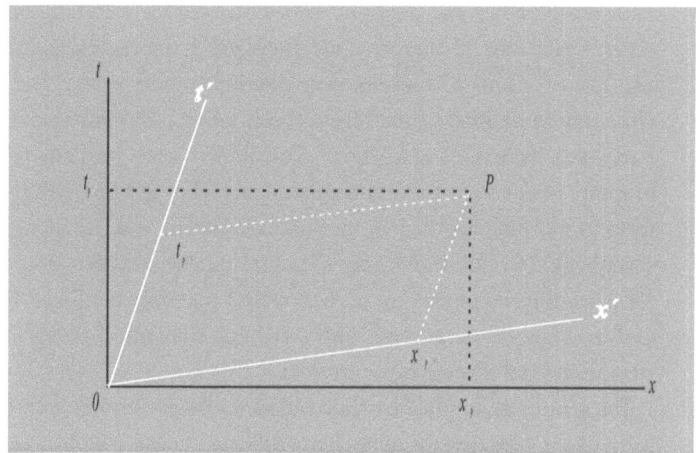

(Figure 6)

What about the axis of x? This is the line that connects all points corresponding to $t = 0$. Any line of the form $t' = $ constant is parallel to this x'axis. But the line $A_1' C_1'$ is just such a line, since A_1' and C_1' are events by which simultaneity in S' is defined. Hence we construct the x'axis by drawing a line parallel to $A_1'C_1'$, and for convenience we make it pass through O, which is thus described both by

$$x = 0, \quad t = 0 \tag{17}$$

and by

$$x' = 0, \quad t' = 0. \tag{18}$$

The noncoincidence of the axes x and x' does not, of course, imply any geometric tilting of one with respect to the other; it is a purely formal tilting in the abstract space constructed from the x and t coordinates.

Now this type of diagram displays for us a key feature of the kinematic transformations of special relativity. In *Fig. 6* any point

343

P in the plane of the diagram represents what is called a *point event*, which can be characterized alternatively by the values of x and t or of x' and t'. And our construction implies that x' and t' alike should be linear functions of both x and t. Similarly, x and t are linear functions of x' and t'. This linearity is a fundamental property of the transformation equations. If they did not have this form, a motion recorded as motion at constant velocity along a straight line in one frame (say S) would not be recorded as uniform rectilinear motion in S'. This would therefore conflict with Galileo's law of inertia and with our basic dynamical condition that all inertial frames are equivalent.

The symmetry implied by the relativity principle means that the form of the relationships must be as follows:

$$x = ax' + bt' \tag{19}$$

with

$$x' = ax - bt. \tag{20}$$

These are set up so as to resemble as closely as possible the Galilean transformation (equations)[433] to which they must certainly reduce for sufficiently small values of the speed v of S' relative to S. The motion of the origin of S as measured in S' is defined by putting

$$x = 0 \tag{21}$$

in the first of these equations. Similarly, the motion of the origin of S' as measured in S is defined by putting

$$x' = 0 \tag{22}$$

in the second equation. The velocities are equal and opposite and both of magnitude v. This gives us the condition

---

433 At this point French refers to the following equations of the Galilean transformation:-- $x = x' + vt$; $x' = x - vt$ which he had quoted earlier in his book.

$$b/a = v \, . \tag{23}$$

Next we consider the descriptions according to S and S' of a light signal travelling in the positive x direction. Let the signal originate at O of *Fig. 5*. It is then described by the following very simple equations in S and S', respectively:

$$x = ct \qquad x' = ct' \tag{24a,b}$$

Substitute these particular expressions for x and x' in equations (19, 20), and we get the following:

$$ct = (ac + b) \, t' \tag{25}$$

$$ct' = (ac - b) \, t \tag{26}$$

Eliminating t and t' between these last equations, and using the condition $b = av$ from equation (25), we find

$$c^2 = a^2 \, (c^2 - v^2) \, . \tag{27}$$

Therefore,

$$\alpha = \frac{1}{(1 - \frac{v^2}{c^2}) \, 1/2} \, . \tag{28}$$

It may be noted that ..."

From this last equation (28) French then deduces the formulae of the Lorentz transformation in a way largely analogous to that Einstein had taken in his original paper on the special theory of relativity[434]

---

434 *Zur Elektrodynamik bewegter Körper* (On the Electrodynamics of Moving Bodies), Annalen der Physik 1905, pp. 891 f.

(2) Let us now look at some formal aspects of French's deduction.

(a) We first consider formulae (17) and (18). In his text French had referred to *Fig. 6,* depicting two reference frames in relative motion. The *x*'axis had been made to pass through 0 'for convenience'. Starting from that premise it was to be understood that $x = 0$ and $x' = 0$.

The space/time coordinates of a stationary reference system (*S*) and a moving system (*S'*) will not coincide permanently but only at a single point, or point event, when the two systems overlap exactly. Formula (17), consisting of two sections separated by a comma, therefore is a mathematically correct description of this single point event; but it is not correct as a description of the *functional* relations between x and 0, nor of those between t' and 0. It is logically correct only if read

$$x = 0 \quad \textit{if, and only if,} \quad t = 0, \tag{29a}$$

which implies

$$\textit{for all } t^1 \ 0, x^1 \ 0. \tag{29b}$$

The same holds for formula (18). It is logically correct only if read

$$x' = 0 \quad \textit{if, and only if,} \quad t' = 0, \tag{30a}$$

which implies

$$\textit{for all } t' \neq 0, x' \neq 0. \tag{30b}$$

(b) We now consider equations (19) and (20). They resemble in some aspects the corresponding equations of the Galilean transformation[435] but

---

435 'Galilean transformation' is the name for a set of equations used for establishing the mathematical relations between a system of coordinates and another system, which is in motion on a straight line relative to it.

French no longer defines the relation between x and x' by velocity (v) and time (t). The new and unknown factors a and b have been introduced; and the time t' of the moving system S' now is a function of distance, described by x and x' and by the unknown factors a and b. It no longer is the same as in S.

That is the point therefore at which Einstein's concept of 'relative time' enters the algebra. In his text below *Figure 6* French mentioned that factors a and b, indicating that the relationships between x and x' and between t and t', are to be linear functions.

(*c*) We now come to equations (21) and (22). In his text referring to those formulae French explains that 'the motion of the origin of S as measured in S' is defined by putting $x = 0$'.

The meaning of that sentence is not clear. We can describe motion in a system of coordinates by a change of the values of space and time. For describing it algebraically we therefore need a functional equation permitting different values of space and time. Hence $x = 0$ cannot be a definition of motion, nor can $x' = 0$.

It seems that French omitted the second part of equations (17) and (18) as explained in formulae (29a,b) and (30a,b) at this point.

(*d*) French's text to formulae (20-23) shows that he deduced from equations (20) and (22) his formula (23), which gives us

$$ax - bt = 0 \quad \text{or} \quad ax = bt \quad \text{or} \quad x = bt/a \qquad \text{(31a,b,c)}$$

In combination with the general functional relationship $v = x/t$ we get $b/a = v$. Formula (21) therefore presupposes the truth of (20) which is not a true description of a functional relationship, but only true for the single point event $t' = 0$. We saw that in (30a,b).

(*e*) It follows from this that the transformations of formulae (25) and (26) to (27) and (28), which presuppose (23), are not based on valid rules of mathematical inference. French's further deduction of the Lorentz transformation rests on those equations.

Index of Names

Index of Subjects

aberration 170

acceleration 44, 57, 69, 151, 154, -and gravity 279-305

action at a distance 8, 31, 36, 165, 167, 279, 283, 289-292

alpha particles 143

alpha rays 142, 148

anthropocentric theory, Kant's 95, 123

anti-Semitism 40

a priori 72ff; Kant's view 98, 103, 104; in Popper's theory 110ff, 115ff; Einstein's rejection of 52, 56, 58, 90,118ff.

astronomy 23, 89, 171f, 282,

atmosphere 137, 139, 176, 180

atom 7, 9, 142ff, 185, 202, 228; bomb 148; internal forces of 37; planetary model of 10, 22, 163, 185ff, 210, 263-278

attraction, force of 8, 35, 37, 174, 277, 279, 283, 288, 290, 308

autonomy, principle of 83, 94-101, 103, 104,109, 110, 115, 123, 320, 323-325, 331, 334-336

axioms, axiomatic method 39, 45, 50-59, 61-82, 83-101, 103-125, 129-135, 145, 157, 158, 193, 196, 201, 203, 207, 214, 219-225, 229, 232-235, 260, 261, 264, 328

barium 143, 144, 147

Big Bang hypothesis 31, 32, 88, 160, 267, 314

binary code 205

birefringence 170

black body 183, 185, 190, 191, 231

black hole 222, 31, 286

Brownian motion 32,

calculation, mathematical 17, 76, 137, 153, 171, 181, 189, 196, 197, 210, 232, 235, 270, 314

calculus 117, 177, 192, 212, 330

catalysers 279, 304

cathode rays 24, 203

causality, principle of 41, 265, 290

clocks 14, 135, 150-155, 226, 228, 239, 242-246, 309

cloud chambers 7, 272, 297,

cognition, process of 50, 83, 91, 92, 96, 124

combustion engine 25

community, scientific 90, 157, 163

computer 22, 66, 79, 205, 319, 327

conditions, ideal 189, 194, 197, 223

conservation, laws of 94, 216, 252, 261

content, informational 52, 53, 66, 74, 79, 122, 202, 205, 218, 247, 290, 326, 327

contradiction, logical 7, 16, 29, 41, 51, 65, 67, 96, 119, 120, 121, 124, 125,140, 162, 208, 212, 215, 218, 225, 228, 233, 235-237, 240, 244-246, 251-253, 265, 270, 273, 284, 285, 288, 296, 337, 340

coordinates 79, 117, 149, 152, 174, 190, 191, 213, 225, 226, 230-232, 234, 237-239, 247, 253, 284-286, 335, 340, 342, 343, 346, 347

Copenhagen Interpretation 32, 34, 41, 183, 187, 188, 210, 211, 327

Copernican revolution, Kant's 83, 94, 122

Copernican theory 95, 166, 276, 279, 295

cosmology 94, 96, 97, 171, 281, 307, 313, 314, 316

cosmos 95, 144, 166, 281

creeds, religious 95, 100, 292

criticism 16, 34, 40, 41, 51, 62, 63, 73,